Social Mobilization, Global Capitalism and Struggles over Food

A Comparative Study of Social Movements

RENATA MOTTA

Institute for Latin American Studies
Freie Universität Berlin, Germany

Routledge
Taylor & Francis Group

LONDON AND NEW YORK

First published 2016 by Routledge

2 Park Square, Milton Park, Abingdon, Oxfordshire OX14 4RN
711 Third Avenue, New York, NY 10017

Routledge is an imprint of the Taylor & Francis Group, an informa business

First issued in paperback 2018

British Library Cataloguing in Publication Data
A catalogue record for this book is available from the British Library

Library of Congress Cataloguing in Publication Data
Names: Motta, Renata, author.
Title: Social mobilization, global capitalism and struggles over food: a
comparative study of social movements / by Renata Motta.
Description: Farnham, Surrey, UK; Burlington, VT: Ashgate, [2016] |
Series: Entangled inequalities: exploring global asymmetries |
Includes bibliographical references and index.
Identifiers: LCCN 2015042293 |
ISBN 9781472479082 (hardback: alk. paper) |
Subjects: LCSH: Agriculture and state–Environmental aspects–South
America. | Transgenic plants–Political aspects–South America. |
Transgenic plants–Social aspects–South America. |
Agriculture and politics–South America. | Agriculture–Economic
aspects–South America. | Social movements–South America.
Classification: LCC HD1857.M67 2016 | DDC 338.1/88–dc23
LC record available at http://lccn.loc.gov/2015042293

ISBN: 978-1-4724-7908-2 (hbk)
ISBN: 978-1-138-35879-9 (pbk)

Typeset in Times New Roman
by Out of House Publishing

SOCIAL MOBILIZATION, GLOBAL CAPITALISM AND STRUGGLES OVER FOOD

This book explores the transformation of Brazil and Argentina into two of the world's largest producers of genetically modified (GM) crops. Systematically comparing their stories in order to explain their paths, differences, ruptures and changes, the author reveals that the emergence of the two nations as leading producers of GM crops cannot be explained by technological superiority of biotechnology; rather, their trajectories are the results of political struggles surrounding agrarian development, in which social movements and the rural poor contested the advancement of biotechnologically based agrarian models, but have been silenced, ignored or demobilized by a network of actors in favour of GM crops.

Based on rich interview and media material collected amongst activists, the author highlights the importance of political struggles over GM crops not only to debates on agrarian futures and food security, but also as illustrations of the challenges faced by contemporary democracies. An international comparative study, this book raises the question of how social mobilization and rights claims can counter the systemic imperatives of global capitalism and political interests, at a time when regional governments are reliant on commodity booms, whilst globally, governments are obliged to introduce programmes of austerity. As such it will appeal to scholars of sociology, political science and geography with interests in social movements, development, globalization, inequality and political economy.

Renata Motta is Assistant Professor of Sociology at the Institute for Latin American Studies at the Freie Universität Berlin, Germany.

Entangled Inequalities: Exploring Global Asymmetries

Series editors:
Sérgio Costa, Freie Universität Berlin, Germany

Departing from classical approaches to the study of social inequalities between individuals and social classes within particular national settings, this series emphasises the production and reproduction of inequalities across borders, as well as the multiplicity of categories – whether 'race', 'sex' or 'nationality' amongst others – according to which contemporary inequalities are shaped.

Entangled Inequalities constitutes a forum and a catalyst for discussing recent advancements in inequality research from a transnational, global and intersectional perspective, highlighting the fact that social inequalities are always the product of both global interpenetrations and of complex intersections between different social categorisations. The series therefore welcomes monographs and edited volumes across the social sciences that deal with inequalities from an 'entangled' perspective – with an intersectional or a transnational focus, or both.

For my parents

Contents

Figures

Tables

Acknowledgements

This project included the support of many people whom I had the pleasure to work with in the last six years as well as from those who have influenced my thinking long before, and from far away. The book had its origins in a PhD project, so I would like to start by thanking Sérgio Costa (mentor, friend and colleague) for being my main exchange partner during this enterprise, and Veronica Schild, who accepted to join the thesis commission when so many decisions had already been taken, and who, nevertheless, was able to help me so much in the crafting of the final piece. The constructive critiques of Manuela Boatcă, Kristina Dietz and Bettina Engels on the thesis defence definitely strengthened the manuscript. During these years in Berlin, I sought feedback from Dieter Rucht, Ruud Koopmans and Wolfgang van den Daele, who gave valuable suggestions for the research. Needless to say, the resulting work is my sole responsibility.

I am indebted to the desiguALdades.net not only for providing me with the financial support for conducting this research but also for the incredible network of people I was able to know and to spend time with during the years in which I had my scholarship there. I thank Marianne Braig, Barbara Göbel, Anna Wickes-Neira, Bettina Schorr, Paul Talcott, Laura Kremmer and Johannes Heeg for maintaining the network. Among many other research fellows, I would like to especially thank Elizabeth Jelin, Juan Pablo Pérez Sáinz, Elisa Reis, Carla Gras and Lena Lavinas, with whom I had the privilege to discuss my chapters. I am grateful to Rodrigo Rodrigues and Manuel Góngora-Mera for their help in the end phase of the manuscript. My colleagues, with whom I shared the office in the beautiful villa in Thielplatz, Maria Fernanda Valdés, Markus Rauchecker, Frank Müller, Constantin Groll and Jairo Baquero, thank you for the mutual support and motivation.

Freie Universität Berlin brought me into contact with great scholars and friends. I thank Sabina García, Olga Piperi, Kristin Wintersteen, Verena Schüren, Christof Mauersberg and Marius Haberland for sharing experiences, thoughts, wishes and plans. I thank my colleagues from the sociology colloquium. The manuscript benefited from the stimulating debates I had with graduate and undergraduate students in the seminars on social movements, political economy and environmental conflicts that I taught in 2014 and 2015 at the Institute for Latin American Studies (LAI). My debt of gratitude also goes to the staff of my colleagues and staff of LAI for the excellent atmosphere that encouraged us to combine research and teaching.

I am indebted to Manuel Bastias and Pablo Holmes, old friends from the Staatsbibliothek zu Berlin, the incredible library where the angels from Wim Wenders' *Himmel Über Berlin* surely looked over our deep theoretical conversations. Far from Berlin but always present in my trajectory these years, Matías Fernandez, I would like to thank you for your care. Other institutions and people were important for this project. The team from the Agricultural Development Unit from United Nations

Economic Commission for Latin America and the Caribbean in Santiago received me for a research stay, and the Dahlem Research School awarded me scholarships for research stays and conference trips. During these six years, I had the opportunity to present pieces of the manuscript at a number of conferences whose participants nurtured and inspired me. I am thankful to the Research Network on Sociology of Risk and Uncertainty from the European Sociological Association and to the Research Committee on Social Movements and Social Classes from the International Sociological Association.

Mediating between the academic world and that of social movements, I was fortunate to meet Carla Poth at desiguALdades.net. Carla became for me the entry door to Argentina. She not only offered me the contacts to all activists I asked for, but she also remained during the whole process my main reader for the Argentinean chapters. Together with Markus Rauchecker, we three formed a group of intense and constructive exchange on Argentina and agrobiotechnology. Other Argentineans who enriched my understanding were Florencia Arancibia, Mauricio Berger and Pablo Lapegna. On the Brazilian side, I am indebted to Gabriel Fernandes, who also introduced to me the world of activism against GMOs. More than that, Gabriel was the archive of the history of the issue in Brazil. He was an important reader and critic of my work. Once again, I hold the entire responsibility for the end result.

I am indebted to Elaini da Silva, Guilherme Leite, Ligia Fabris and Florencia Arancibia for their invaluable efforts to edit and revise the final versions of the manuscript. The constructive suggestions from the anonymous reviewer for Routledge surely improved the manuscript. My special gratitude goes for Neil Jordan, Senior Commissioning Editor at Routledge, for his engagement with the project. I thank Forrest Kilimnik for his extraordinary and meticulous copyediting work.

My family and friends spread around this world know how much I am grateful for having their love and support throughout this time of my life. My deepest debt of gratitude is towards Nicolás Carosio. With his love and care, I feel that all projects can be accomplished. Our just born child had a crucial role in this project by announcing a different project on the way that would demand all my attention and care.

Last but not least, my deepest gratitude goes to all activists and social movement leaders who spent their precious time with me and told me their story, as part of the commitment to their ideals and struggles.

Abbreviations

AAPRESID:	Asociación Argentina de Productores en Siembra Directa (Argentinean No-Till Farmers Association)
ABIA:	Associação Brasileira das Indústrias da Alimentação (Brazilian Association of Food Industry)
ABIOVE:	Associação Brasileira das Indústrias de Óleos Vegetais (Brazilian Association of Vegetable Oil Industries)
ABRASEM:	Associação Brasileira de Sementes e Mudas (Brazilian Association of Seeds and Seedlings)
ANVISA:	Agência Nacional de Vigilância Sanitária (Brazilian Health Surveillance Agency)
AS-PTA:	Assessoria e Serviços a Projetos em Agricultura Alternativa (Advisory and Services for Projects in Alternative Agriculture)
BRASPOV:	Organização Brasileira de Obtentores Vegetais (Brazilian Organization of Plant Breeders)
CASAFE:	Cámara de Sanidad Agropecuaria y Fertilizantes (Chamber of Plant Health and Fertilizers)
CCT:	conditioned cash transfer
CDB:	Convention on Biological Diversity
CETAAR:	Centro de Estudios sobre Tecnologías Apropiadas de Argentina (Centre for Studies on Appropriate Technologies in Argentina)
CNA:	Confederação Nacional da Agricultura (National Confederation of Agriculture)
CNBS:	Conselho Nacional de Biossegurança (National Biosafety Council)
CONABiA:	Comisión Nacional Asesora de Biotecnología Agropecuaria (National Commission of Agrarian Biotechnology)
CONICET:	Consejo Nacional de Investigaciones Científicas y Técnicas (National Council for Scientific and Technical Research)
CONINAGRO:	Confederación Intercooperativa Agropecuaria Limitada (Federation of Cooperatives)
CONTAG:	Confederação Nacional dos Trabalhadores na Agricultura (National Confederation of Workers in Agriculture)
COP-MOP:	Conference of the Parties/Meeting of the Parties
CPT:	Comissão Pastoral da Terra (Pastoral Land Commission)
CRA:	Confederaciones Rurales Argentinas (Argentinean Rural Confederation)
CTA:	Central de Trabajadores Argentinos (Argentine Workers' Central Union)

CTNBio:	Comissão Técnica Nacional de Biossegurança (National Technical Commission on Biosafety)
CUT:	Central Única dos Trabalhadores (Central Workers' Union)
Embrapa:	Empresa Brasileira de Pesquisa Agropecuária (Brazilian Corporation of Agriculture Research)
EU:	European Union
FAA:	Federaciones Agrarias Argentinas (Argentine Agrarian Federation)
FASE:	Federação de Órgãos para Assistência Social e Educacional (Federation of Organs for Social and Educational Assistance)
FETAG:	Federação dos Trabalhadores na Agricultura (Agricultural Workers Federation)
G-77:	Group of 77
GDP:	gross domestic product
GM:	genetically modified
GMO:	genetically modified organism
GRAIN:	Genetic Resources Action International
GRR:	Grupo de Reflexión Rural (Group of Rural Reflexion)
IBAMA:	Instituto Brasileiro do Meio Ambiente e dos Recursos Naturais Renováveis (Brazilian Institute of Environment and Renewable Resources)
Ibope:	Instituto Brasileiro de Opinião Pública e Estatística (Brazilian Institute of Public Opinion and Statistics)
IDEC:	Instituto Brasileiro de Defesa do Consumidor (Brazilian Institute for Consumer Protection)
IMF:	International Monetary Fund
INTA:	Instituto Nacional de Tecnología Agropecuaria (National Institute of Agricultural Technology)
MCC:	Movimiento Campesino de Córdoba (Córdoba Peasant Movement)
MMC:	Movimento de Mulheres Camponesas (Movement of Peasant Women)
MML:	Movimiento de Mujeres Agropecuarias en Lucha (Movement of Agrarian Women in Fight)
MNCI:	Movimiento Nacional Campesino Indígena (National Peasant Indigenous Movement)
MOCASE-VC:	Movimiento Campesino de Santiago de Estero-Vía Campesina (Santiago de Estero Peasant Movement – Vía Campesina)
MPA:	Movimento de Pequenos Agricultores (Movement of Small Farmers)
MST:	Movimento dos Trabalhadores Rurais Sem Terra (Landless Workers' Movement)
NGO:	non-governmental organization
NIMBY:	not in my backyard
OBC:	Organização Brasileira de Cooperativas (Brazilian Organization of Cooperatives)
PJ:	Partido Justicialista (Justicialist Party)

PMDB:	Partido do Movimento Democrático Brasileiro (Brazilian Democratic Movement Party)
PRONAF:	Programa Nacional de Fortalecimento da Agricultura Familiar (National Program for Strengthening Family Farming)
PSB:	Partido Socialista Brasileiro (Brazilian Socialist Party)
PSDB:	Partido da Social Democracia Brasileira (Brazilian Social Democracy Party)
PSOL:	Partido Socialismo e Liberdade (Socialism and Freedom Party)
PT:	Partido dos Trabalhadores (Workers' Party)
PV:	Partido Verde (Green Party)
RR:	Roudup Ready
SBPC:	Sociedade Brasileira para o Progresso da Ciência (Brazilian Society for the Advancement of Science)
SENASA:	Servicio Nacional de Sanidad y Calidad Agroalimentaria (National Service for Agricultural Health and Quality)
SERCUPO:	Servicio a la Cultura Popular (Service to the Popular Culture)
SPS Agreement:	Agreement on the Application of Sanitary and Phytosanitary Measures
SRA:	Sociedad Rural Argentina (Argentinean Rural Society)
SRB:	Sociedade Rural Brasileira (Brazilian Rural Society)
SSDRAF:	Subsecretaría de Desarrollo Rural y Agricultura Familiar (Under-Secretary of Rural Development and Family Farming)
STS:	science and technology studies
TRIPS:	Agreement on Trade-Related Aspects of Intellectual Property Rights
UPOV:	Union Internationale pour la Protection des Obtentions Végétales (International Union for the Protection of New Varieties of Plants)
US/USA:	United States of America
WTO:	World Trade Organization

Introduction

Brazil and Argentina are the second and the third largest producers, respectively, of genetically modified (GM) crops.[1] While Argentina had been for a long time producing more than Brazil, in recent years Brazil has exceeded its neighbour. Their stories are deeply connected. The first GM seeds planted in Brazil were smuggled from Argentina by farmers along the southern border who were impressed by what they heard from their neighbours regarding the performance of the new technology. That GM soy that arrived illegally from Argentina became known in Brazil as *soja Maradona*, in reference to the legendary Argentinean soccer player and its performance. In 2012, a new type of genetically modified soy was first launched in Brazil. The technology owner, the US-based multinational biochemical company Monsanto, threatened not to launch its new GM soy in the Argentinean market as long as the country did not reach an agreement that protected the company's intellectual property rights over the seeds and its uses, thereby guaranteeing a system of royalty collection as there is in Brazil. In 2014, with an agreement being reached, the Argentinean secretary of agriculture announced that Monsanto would launch a new variety in the country. He went further to state that Argentina did not want its farmers to have to smuggle the *soja Ronaldinho* (Camandone 2014).

However, this analogy can be misleading. The story behind the transformation of these two countries into the world's largest producers of GM crops cannot be reduced to technological performance being equated to the talent of a soccer player. Just as in soccer, the playing field in agriculture has changed its very structure. In soccer today, the team that wins is not necessarily the one with the best player but more often the team better equipped with resources to compete globally for the talented players and coaches. Soccer teams that already possess more resources and better political connections are favoured in this new context together with those teams located in countries with the best infrastructure and public policies supporting sports. Uniforms, stadiums and championships have been transformed into the most expensive publicity spots for world brands, with private television channels buying exclusive rights for transmitting events. All possibilities of opening new markets have been explored. Stadiums have been privatized and transformed into exclusive places where tickets are not accessible for the masses. All these processes of commodification of soccer were accompanied by increasing levels of corruption and lack of transparency. The state has little control over these processes because a private organization is in charge of regulating soccer internationally, while countries hosting world championships often abide by the rules of transnational corporations dominating the soccer business, thereby suspending constitutional norms and rights protection.

The analogy is not perfect but suggests many parallels to the context in which *soja Maradona* and *soja Ronaldinho* came into existence. The introduction of biotechnology is part of a global process of structural change in agriculture and the food industry, which includes, among other factors, an accelerated process of commodification of seeds and land, the increased integration of global commodity chains, an intensification of food processing and stronger presence of industrial foods in food habits, as well as the corporate control in all nodes of the food chain, that is to say, from farm to fork. This material basis of contemporary global agrarian markets receives symbolic support in the negotiation of international legal instruments that construct a policy culture in which science and risk are harmonized as criteria for decision-making. The regulation of biotechnology is thus situated in a context of innovation policy, which assumes the beneficial aspects of innovation and that the state should not create obstacles to it. To the contrary, the state promotes agrobiotechnology through agrarian policies that open up new markets, thus commodifying agrarian production from input to the final product. Examples of commodification processes in agriculture are intellectual property rights over seeds, financialization of agrarian production and lack of oversight over land entitlements in favour of the land market. The state promotes technologies and infrastructure for large-scale production integrated into world markets, ensuring legal and financial conditions for it. The state also neglects the socio-environmental effects of these changes that violate constitutional and human rights.

This global structuration of agrarian markets increases the gap between two models of agrarian development: (1) agribusiness-producing commodities for export versus (2) small-scale production of food for local and regional consumption, often based on agroecology. While the state subsidizes and supports the former, the latter is left to its own strategies. This means an asymmetrical distribution of wealth, rights and risks that in turn create new inequalities. The political decision for promoting agrobiotechnology reproduces the socioeconomic inequalities of a capital-intensive technological package that demands large-scale production to generate profits and concentrate rents. These economic inequalities are translated into the political system, which more represents the interests of capitalist farming. The approval of GM crops relies on asymmetries of knowledge in regulating them, which downplay the risks involved as well as an asymmetrical distribution of health and environmental risks and damage. The rural poor in export countries suffer the biggest burden of the negative consequences of an expansive, profitable and chemical-intensive commodity production: they suffer violations of their right to land, to their cultural traditions and to a sound environment and a healthy life.

For these reasons, such global processes of agrarian change might trigger social mobilization to protest against the violation and diminishing of their rights. GM crops provide a good entry point to discuss the model of agrarian development that such changes represent. Indeed, global market dynamics and state policies have not managed to avoid social mobilization and resistance. There are counter-movements that dispute the thinking that agriculture can be entirely commodified. A political struggle over agrarian futures is taking place at global, national and local levels. In addition, global processes of agrarian change do not follow the same pattern everywhere, with innovation travelling from a centre to the rest of the world. As they are mediated by politics, and influenced by each context, different models and different futures are possible.

The fight against GM crops and for a different model of agrarian development is also a dispute within the state, in its complexity due to the existence of different actors, powers and institutions, each with specific aims and constituencies. States are the main targets of struggles over genetically modified organisms (GMOs) because all sides make demands to the state. On the one side, markets and global economic actors need legal reassurances for their investments and make claims to state authorities, demanding the state to guarantee an adequate institutional framework for their investments. On the other side, citizens demand that state authorities protect and promote their constitutional rights. Notwithstanding the influence of transnational actors and ideas, processes of legitimizing a given policy take place at the state level (Costa 2006). States not only remain the main guarantors of citizenship and rights but also have the decision-making authority to approve or prohibit GM crops in their territory. This means that there is some – asymmetrically distributed – room for autonomy in the manner that individual states negotiate and interpret such global trends when deciding which role biotechnology should take in their development strategies. Due to such relative autonomy in establishing policies as instruments of country development, market 'regulation' and rights promotion, state responses vary, with some embracing GMOs and others rejecting or strictly regulating them. Indeed, Argentina and Brazil show varied responses to the launching of the new technology.

Two contingent stories

In 1996, Argentina was a pioneer in adopting GM crops, together with the USA, where the technology holder, Monsanto, is based. The agricultural authorities approved the new technology in a bureaucratically insulated way, without allowing the issue to be part of the public agenda. This resulted in a rapid conversion to a new model of agriculture anchored in transgenic soy also known as sojazation (*sojización*). In an influential article published in a national leading academic journal, *Desarrollo Económico*, three known economists, Bisang, Anlló and Campi (2008), from the United Nations Economic Commission of Latin America and the Caribbean (ECLAC), have described and named the process as 'a (not so) silent revolution'. Their argument is that a fundamental and very visible change was under way in the mode of production in the Argentinean agrarian sector. Since the 1990s, the country had been experiencing a transition from a model in which landholders also farm the land, using their own machinery and traditional expertise and know-how, to a new model, in which agribusiness firms coordinate agricultural activities without owning the means of production (land and machinery) and rather by subcontracting services while relying on a different type of knowledge, namely management and entrepreneurship. These companies have devised a complex network and contractual relations to engage in this business. Indeed, admirers and critics alike speak of a revolution or a new model of agriculture in the country (Gras 2012; Gras and Hernández 2008; Pengue 2005; Teubal and Rodríguez 2002; Varesi 2010). This only gives more prominence to the adjective chosen by Bisang, Anlló and Campi: silent. They do not elaborate further on the use of this term; however, it hints at an important aspect of the so-called revolution in Argentina.

There is also another side of the story, from the point of view of actors who were affected by the advancement of the revolution and who had no chance to participate in the shaping of the new model. Their claims were dismissed or, worse, silenced. The agrarian change in Argentina constituted what actors call the agrarian model (*el modelo agrario*), the soy model (*el modelo sojero*) or simply the model (*el modelo*). The narrative of social mobilization surrounding this process in Argentina is told through a wide range of actors that fall into three social movements categories: the environmental movement, the agroecology movement and the peasant movement. In Argentina, groups of neighbours directly affected by the widespread adoption of GM crops have organized themselves and become important collective actors in that fight. Scientists and university teachers, journalists and doctors are important allies and work with social movements. There has neither been an actual campaign nor a coalition against GM crops; rather, movements have been fighting different and parallel struggles. One point of convergence is their agreement on the diagnosis of the source of the problems, namely the model of agrarian development in Argentina anchored in the technological package composed of GM seeds and pesticides. This is identified as a common enemy to their campaigns concerning deforestation, peasants' rights, and health and environmental protection from pesticides. Biotechnology itself has hardly been at the centre of their claims; instead, it has been considered a component of the cause, with divergent degrees of importance in the causal chain. At times, these parallel fights meet and converge, forming alliances, and other times, continue to run parallel.

Brazil, today the second world producer and exporter of GM crops, might also be in the top ranking of world social mobilization against GM crops. How can this be interpreted? Indeed, the story of GM crops in Brazil is far from linear. Expecting that the country would follow the relatively unchallenged script that guided the technology adoption in the pioneer countries, the proponents of GM crops were faced instead with sharp opposition from civil society organizations and social movements, and also from subnational governments. In the late 1990s, a long dispute over Brazilian policy for agrobiotechnology began. Reality and legality were often at odds: GM seeds were smuggled from Argentina and illegally cropped. This paved the way for what activists have called politics of *fait accompli*. Huge battles to shape the law and its application ensued. From an ambivalent position, the governmental policy has turned into a clearly pro-GMO position; nonetheless, political mobilization against the technology endures. This long and complex dispute involves different levels of government, rifts inside the governing party and its coalition in power, conflicts between the executive and the judiciary power, the mobilization of the parliament, as well as different social movements.

The model of agrarian development that was consolidated in Brazil is known as agribusiness (*agronegócio*). Due to a strong mobilized peasantry in the 1990s, disputes over the agrarian policy preceded the arrival of GM seeds. Biotechnology has provided an entry point for alliances with urban movements in the resistance against the neoliberal restructuring of agriculture in the direction of agribusiness. Different social movement sectors were involved in the social mobilization against GMOs in Brazil: the peasant movement, the environmental movement, the agroecology movement, the consumer rights movement and the human rights movement. Moreover, scientists, students and universities were important allies and worked with social movements. Given the diversity of their history and claims, each of these movement sectors incorporated the fight against GMOs in their past struggles in different ways. Their framing activities, resources mobilized and forms of action have been specific to their constituencies but nevertheless converging in a common goal. This is reflected in the name of their joint campaign: GM-Free

Brazil. Under this heading, this variety of movements had very different yet complementary and reinforcing demands.

The comparison between Argentina and Brazil makes sense because, in principle, they could have experienced similar trajectories. It could have been the case of a successful and smooth introduction of GMOs, following the plans of biotech corporations and agrarian actors and policy-makers in Brazil, as their position indeed prevailed after the first ten controversial years. Conversely, there could have been a disputed transformation in Argentina, as increasingly became the case later. In sum, behind the current convergence of Argentinean and Brazilian trajectories in the world ranking of GM crops, these are the results of very different paths of contentious politics. This contingency in how the history of GMOs evolved in these countries makes a comparison between them interesting and fruitful for an exploratory research on the causes that explain each path.

Already the research undertaken in both countries shows striking differences on the analytical focus. Argentina is considered a typical case of bio-hegemony by Newell (2009), and the networks that installed and promoted this model have been a subject of ethnographic research (Hernández 2007; Córdoba 2013). The understanding of what became known as the biotechnological model, the transgenic revolution, the new agrarian model and agribusiness prevails. These studies are divided between enthusiasts (Bisang *et al.* 2008; Trigo *et al.* 2002) and critics (Cáceres 2015; Giarracca *et al.* 2001; Gras 2009; Gras and Hernández 2008, 2013; Leguizamón 2014; Pengue 2005; Poth 2013; Teubal and Rodríguez 2002). Critics have emphasized the inequalities in the distribution of rents, lands, risks and ecological burdens. There is recent research on social resistances that disputes the model hegemony (Delvenne *et al.* 2013), focused on the environmental and health consequences of pesticides used in the technological package (Arancibia 2013; Carrizo and Berger 2009), or on forced evictions with the expansion of GM soy fields (Lapegna 2013a).

In Brazil, by contrast, there is more specific research on GMOs as a controversial issue. It became a case for scholars from science and technology studies, looking at technoscientific controversies (Cesarino 2006; Guivant 2006; Guivant and Macnaghten 2011; Ninis 2011), as well as studies on the role of experts and of public debate in a democracy, revealing the disputes for legitimacy in the construction of social representations of GMOs (Camara 2011; Leite 2007; Lima 2007; Menasche 2003, 2005; Ribeiro and Marin 2012; Santos 2007; Silveira 2004; Silveira and Almeida 2008). It also provided plenty of material for social movement scholars, looking at different movement sectors and strategies (Bissoli 2012; Freitas 2011), including resistance on a small scale in searching for alternatives (Reis 2012). Brazil has also been studied from the analytical perspective of networks of actors and their strategies to establish material, institutional and discursive power in a project to convert Brazil to biotechnology (Castro 2006; Lisboa 2007, 2009; Pelaez and Schmidt 2000), and the political economy and structural conditions for that (Benthien 2010).

The present book expands upon existing research and makes an original contribution by undertaking a comparative study of Brazil and Argentina, a case of successful (at a first stage) social mobilization and a case of absence thereof. It explores the differences in the paths of adoption of GM crops that are strongly related to differences in the processes of social mobilization while also tracing and explaining the changes over time. Whereas most existing works choose to focus on one type of movement, this book investigates 25 years of the contested transformation processes experienced by these countries by interviewing activists from all movement sectors involved in the issue. Consequently, the book covers a wide array of topics that social movements have raised: from scientific debates to agrarian policy, as well as consumer rights and legal debates, bridging urban-based struggles and mobilizations on the countryside.

The argument

The ten-year lag between Argentinean and Brazilian conversion of the majority of their soy fields to GM soy, in 1999 and 2009 respectively, shows how politics in general, and social mobilization in particular, influenced the trajectory of the technology in each case. Whereas the coalition of actors promoting biotechnology was able to portray GM seeds as being beneficial to the society at large and avoid public debate on its disadvantages, in Brazil social movements were successful in disputing the meanings associated with GMOs and in making it a controversial technology. However, that success did not last forever, and was later overthrown by the pro-GMO coalition. The reconstruction of the Brazilian history shows how the dominance of agribusiness and biotechnology depends on the engagement of a variety of actors in the pro-GMO coalition, as in Argentina. This means that although social mobilization can dispute and shape the course of agrarian change, the role of politics in the trajectories of societal change is heavily influenced by the structural inequalities that characterize the context of the dispute.

In short, the argument developed in these pages is that the trajectories of adopting GM crops are stories of political struggles among different actors, in which the playing field has been crucial. The transformation of Argentina and Brazil as top world producers of GM crops for export is not an automatic result of market dynamics. The dominance of GM crops in the countries studied did not follow from technological and economic efficiency (Pelaez and Schmidt 2000) but from the strategies of interested actors to ensure the conditions for a flourishing agribusiness. These included the use of illegal means and the suppression of contestation when necessary to avoid public debate and democratic participation. Moreover, the dominance of GM crops occurred at the cost of land expropriation, environmental degradation, environmental contamination, health damage, and disregard for the law and democratic rights. Such strategies were favoured by a specific context in the political economy in these countries, in which the production of commodities acquired a renewed importance for state revenues, and in which economic growth and increase of consumption had strong legitimizing effects. All in all, it is a result from political struggles on agrarian development. Behind similar positions of market leaders in GMO production, there are very different trajectories revealing the role of state intervention, market strategies and social mobilization. The study assumes that without protest and social contestation, such processes would lead to more inequalities and diminishing of rights.

In both countries, there is enough evidence of the success of creating an agrarian model characterized by corporate domination, free trade orientation and intensive use of biotechnology. It seems to be consensual or even hegemonic. The research identifies attempts to challenge this process by social movements as well as the reactions from authorities, experts and economic actors for suppressing dissent, silencing claims and demobilizing resistance. By examining explanations for the different paths from Argentina and Brazil, I am concerned with understanding both the conditions in which challengers from social movements and civil society organizations changed the official pro-GMO policy as well as the conditions that suppress and silent dissent. This situates the research problem as the more general issue of how contestation on the part of mobilized and affected citizens can counter global processes of creating an integrated and asymmetrical world system of food production and consumption. The wider research problem will be addressed by means of more concrete and specific research questions. A first group of questions concerns the mechanisms of mobilization and demobilization as well as the factors that explain the different outcomes and the change in outcomes over time. The questions are why did Brazilian social movements achieve a moratorium on GMOs and create a controversy about the new technology? Why could Argentineans not shape a similar path? Why did contentious Brazilians lose the battle? Why did Argentineans, although mobilizing, not challenge bio-hegemony?

Among the conditions that might influence social movements' struggles, I am interested not only in looking at national political contexts but also in understanding these in relation to global capitalism and, in particular, its dynamics in agriculture and commodities. Last but not least, investigating the politics of GMOs in Argentina and Brazil is even more interesting for the research problem, as these countries have elected, at the beginning of the twenty-first century, left or centre-left governments, with an agenda of addressing and reversing the effects of neoliberal structural adjustment policies from the 1990s. This begs the question of how the new governments have differed from forerunners on the issue at hand and how the changes in the political economy affected the contentious politics over GMOs. Therefore, a second group of questions inquire into how the structural location of these countries in an asymmetric world agrarian system affected the disputes: How did a new context in the political economy affect movements' opportunities in Argentina and Brazil? How did the integration of GM crops in the production node of global commodity chains affect social movements' opportunities?

A dialogue between political economy, social movements and peasant studies

The book concentrates on the goal of understanding the conditions that allowed social mobilization to shape the trajectories of GMOs and the conditions that were responsible for preventing it from doing so. Central for addressing the book's aim are the political economy and the political ecology of the production of GM crops. Conflicts over distribution of wealth, over the distribution of socio-environmental damage, as well as over access and control of natural resources permeate these stories. By tracing changes over time, the book compares the policies of neoliberal structural adjustments in the 1990s with the 'left turn' since 2003, which is associated with a stronger state that uses commodity exports as 'an opportunity' to reduce poverty. A second field is the policy culture that restricts scrutiny to scientifically demonstrated health and environmental risks. A third topic is the role of media and public debate for the democratic opening or closure of the decisions regarding agrarian development. In addition to such macro-sociological themes, the book identifies and investigates actors and their interactions. On the one hand, there are networks comprising multinational corporations in the seed and chemical industry, agrarian elites, scientists and policy-makers, which deploy various strategies to promote biotechnology; on the other hand, there are social movements and peasant mobilization, which are the main actors contesting it.

Based on the argument that social mobilization is the main mechanism to explain the different trajectories in the adoption of GMOs in Argentina and Brazil, the main data, analytical focus and conceptual framework that guide this study are affiliated with social movement research. By drawing on social movement theory and cumulative research findings, this study attempts to explain why social mobilization against GM crops emerges, how social movements interact with opponents and authorities, as well as what are their forms of action and their impacts. Despite the focus on the social movements, the research takes into account the relations between these actors with other actors disputing the politics of GMOs, including the structural dynamics that influence such disputes, specifically the political economy of food and agriculture.

Therefore, the book combines macro-explanations with the study of actors and their interactions. Following the lessons learned through previous research, it starts with the assumption that many factors are necessary to explain the politics over GMOs: (1) biotechnology is a component of the world agrifood system, in which countries are asymmetrically positioned to influence its capitalist dynamics; (2) institutional support for GMOs is guaranteed by a policy culture that restricts scrutiny to scientifically demonstrated health and environmental risks; and (3) a network comprising multinational corporations in the seed and chemical industry, agrarian elites, scientists and policy-makers deploy various strategies to promote biotechnology. These three factors, when combined, create a hegemonic conception that biotechnology means progress in agriculture and serves societal general interests. The study assumes that, in contrast, (4) public debate and democratic opening of the decisions regarding agrarian development have the potential to challenge that hegemonic conception, and (5) social movements and peasant mobilization are the main actors demanding such an opening.

Setting these factors in relation, the theoretical argument establishes a dialogue between concepts derived from political economy and political sociology. From the former, the concept of food regime (Friedmann and McMichael 1989) will be used to characterize the structural context in which biotechnology appears, connecting it to the agrarian bases of contemporary world capitalism. The materiality of these changes depends, in turn, on specific institutions and also on a symbolic order. Therefore, another key concept from political economy used for this research is bio-hegemony (Newell 2009). It defines the symbolic domination that is in place when the actors that most benefit from GM crops are able to portray GMOs as representing the general public interest. Such a Gramscian approach that underlines power asymmetries in the construction of a hegemonic order (Gramsci 1971) will be complemented with a Habermasian approach, which calls attention to the conditions under which processes of the decision-making and public opinion formation might be considered legitimate (Calhoun 1992; Habermas 1992, 2008). Indeed, civil society actors engage in politics due to their normative expectations regarding democratic politics. While Habermas emphasizes the structural conditions for deliberative politics to take place, this study focuses on the actors that attempt to close public debate or, conversely, to open it. Therefore, from political sociology, social movement studies provide the main concepts for analysing the interaction among contending actors. Additionally, peasant studies provide concepts to understand other forms of politics and reactions among the rural poor affected by the expansion of GM crops. Ultimately, it is in their resistance or accommodation that situations of domination or contestation can be observed.

By bringing literature of political economy and social movements together, this study reacts both to the renewed interest among social movement scholars in bringing capitalism back into their analysis (Barker *et al.* 2013; Stanley and Goodwin 2013) as well as to the call of taking politics seriously to construct critical scholarship on processes of agrarian transformation (Borras 2009). While much focus has been placed on the political structures influencing the possibilities of movements to act, less has been said about how the interplay between politics and markets has affected the campaigns of social movements. Stanley and Goodwin (2013) suggest to give attention again to the enabling or constraining effects of the dynamics of capitalism in collective action, including its indirect and long-term effects. They note the disappearance of capitalism at least in English-speaking scholarship; however, social movement scholars in Latin America have claimed that capitalism could never be neglected in the analysis of social movements in the region (Shefner 2004). Particularly interesting in the present research is to situate the contentious politics over GMOs in the world agrifood system in which Argentina and Brazil are exporters of agrarian commodities and investigate how this position in global capitalism influences movements' campaigns against agrobiotechnology. Although researchers (Schurman and Munro 2009) have investigated strategies of activists to target capital as well as how the insertion of GM products in a global commodity chain have meant more entry points of activism, there has been no critical assessment of how such structures of production constrain activism.

Writing on the challenges involved with rethinking contemporary agrarian change, Borras (2009) argues for the need of critical theories – in order to address the persistent rural poverty and increasing inequality in the countryside – to analyse the politics of agrarian change. Borras states that a critical theory must entail a relational perspective on rural development in contrast to residual theories. The latter contend that rural poverty is caused by the market exclusion of the rural poor and thus will be solved by including them. A relational perspective, however, situates the cause of rural poverty in the 'the very terms of poor people's insertion into particular patterns of social relations; the solutions therefore are transformative policies and political processes that restructure such social relations' (Borras 2009, 13). He advocates an analytical framework centred on state-society relations that avoids one-sided analysis of state institutions or of social movements and social classes, while focusing on their interactions (Borras 2009, 21). Such a proposition converges to a great extent with social movement studies.

This is especially interesting in the case of GMOs because social scientists who are known for their pro-GMO positions rely on the residual theories for rural development by claiming that rural poverty will be solved by bringing the new technological package to peasants through a new 'Green Revolution' (Herring 2007; Paarlberg 2013). They also blame social movements and, in particular, non-governmental organizations (NGOs) based in the Global North that oppose GMOs for denying poor peasants in the Global South the benefits of biotechnology. A relational perspective, by contrast, brings into the centre of analysis the political process in which specific actors make claims on the state to influence policy-making according to their interests. In the case at hand, a coalition of actors that defend the model of large-scale capitalist farming are the main proponents of GMOs, whereas local and national social movements claim alternative models of agrarian development. Looking at their organizational bases, the meanings and the values that they attach to their farming practices, as well as the demands they make to the state, the argument exposed by proponents of GMOs that the new technology benefits small farmers' interests becomes less plausible. It assumes that peasants do not know what is good for them.

At the same time, the themes, actors and settings that are objects of studies on rural development and agrarian change might bring new light to the studies on social movements, which are mostly focused on urban settings. Biotechnology provides an excellent topic for such a dialogue. Researchers of food regimes (Pechlaner and Otero 2008) have argued that linking opposition to GMOs and protests against neoliberal agrarian policies is fundamental if civil society is to participate in shaping the trajectories of food regimes. Due to the strong impact of neoliberal policies on the agrarian structures of developing countries, the rural poor are more likely to resist and to mobilize against the food regime. Their mobilization potential among grass-roots bases is necessary if contestation over GMOs is not to remain restricted to some NGOs and urban activists – as happened in the United States of America (USA) (Schurman and Munro 2010) – and, instead, reach the agenda of mass movements.

The rural poor[2] have increasingly organized themselves around the collective identity as 'peasant movements' and have been amongst the most active movement sectors in the last decades (Borras *et al.* 2008), as seen in the work from the transnational grass-roots social movement organization Via Campesina (Desmarais 2007; Edelman and James 2011; Martínez-Torres and Rosset 2010). Peasants constitute the radical trend of what Holt Giménez and Shattuck (2011) call 'global food movements', as they are structurally combative against the actual system. The global food crisis in 2008 has initiated a debate on the reform of the corporate food regime, in which global food movements have increasingly participated. The promotion of GMOs remains central to the agenda of the hegemonic actors, whereas the rejection of GMOs is a non-negotiable position of the radical movements (Holt Giménez and Shattuck 2011).

In short, social movements and peasants are the main actors posing the political question of alternatives to global capitalist dynamics that seem to control democracy and exacerbate global, national and local inequalities. At the same time, peasants and farmers are also the ones who will define in their everyday lives whether they will follow capitalist discourses and aspire to become small capitalists, expecting that by adopting GM crops they will be able to accumulate wealth and grow. In the words of some peasant leaders, this means that they will be doing small-scale agribusiness (*agronegocinho*). The rural poor, in general, might also cope with the advancement of GM fields through a variety of forms other than social mobilization. This might take the form of 'everyday peasant politics' (Kerkvliet 2009; Scott 1987) as the rural poor resist the effect of capitalist change in their everyday lives in mostly invisible forms, or even get used to living surrounded by GM crops (Lapegna 2013b, Lapegna 2014).

This leads to the incorporation of another lesson learned through past research, namely looking at multiple levels of the contentious politics over GMOs: global, transnational, national and local. A lot has been said about the global scale of agrarian transformations; the role of transnational factors, such as networks of actors and cross-border diffusion of meanings; and the key role of national politics due to the state prerogatives in defining GM policy. At the same time, as with other pressing issues in the global agenda – such as climate change – the effects are tangible and real at the local level although the causes are global and the responsibilities are asymmetrically distributed. They are felt at the level of the bodies (Berger 2013): in racialized ways, gendered ways and class-divided ways. With GM crops, the concentration of socio-environmental effects at the production nodes of the chain implies that the local level of political action is particularly relevant for contesting bio-hegemony or accommodating it. All sides disputing the fate of GMOs move across scales of action.

The structure of the book

Following this introduction, in Chapter 1, the considerations above are brought together in a literature review while identifying the main theories and concepts that have been applied to explain why GMOs became a subject of social disputes and controversy. Building on this review, the relevant concepts for this study are then presented and explained. Next, these concepts are analytically related in a theoretical argument. The chapter further presents the methodological decisions on how to approach the research problem, including the research design, case selection, timeframe and data sources. In order to understand the role of the social mobilization in the process of adoption of GMOs in these relatively similar two countries, a comparison, which acts as a very promising research design for explorative purposes, is made between the different trajectories. In addition to comparing two countries, there is a comparison within the cases concerning time in order to account for the changes in the political economy context and how these transformed the political struggles between proponents of and opponents to GM crops. The chapter then presents the different types of data used in the research: official databanks, documental sources, secondary literature and primary data collected in interviews, in newspaper articles and on the Internet. The main empirical data is based on in-depth interviews conducted with key activists and social movement organizations during the years 2012 and 2013.

Chapters 2 through 5 are the empirical core of the work. The first two chapters reconstruct the Argentinean case. Chapter 2 presents the context in which GM crops were approved and introduced in Argentinean fields, a historical moment deeply marked by the neoliberal structural adjustments taking place in the country. The traditional civil society organizations, such as workers' unions, were demobilized by the very mechanisms of the model. The 'transgenic revolution' engaged state officials, experts, agrarian elites and chemical companies in the construction of a shared meaning that defined GMOs as promoting the welfare of Argentina and its citizens. The chapter includes the views of social actors who suffered the negative and direct consequences of the advancement of GM soy and also from actors who reacted to biotechnology. However, social mobilization disputing this meaning had no resonance in the wider public and political debate. The watershed is the financial crisis that started in December 2001. It not only changed the overall context for any social movement action in the country but also erupted in a cycle of contention. GM crops for export played a pivotal role in the recovery of the crisis.

Chapter 3 analyses social contestation over the biotech food regime during *kirchnerismo* (2003–). It is divided into two parts. The first part starts with the new presidential office, when social movements had to interpret the structure of political opportunities and look, without success, for spaces of influence. It ends in 2008 when a polarized conflict between agrarian elites and the government took over the national agenda for some months and made clear the political alignments between the agrarian sector and the government regarding the policy on genetically modified organisms. The conflict changed the context for movements to act, as they did not believe in the opportunity of achieving changes at the national level. Mobilization, though, increased by targeting other scales while activists started to create new political spaces at the local level, while also targeting the global and transnational arenas. Therefore, this chapter tells the story of how bio-hegemony was sustained while becoming increasingly an object of social mobilization. From a (not so) silent 'transgenic revolution' as narrated in the previous chapter, the next phase shows much more 'noise' over the agrarian model, even though it has not resonated in the Argentinean national political system.

Chapters 4 and 5 are reserved for the Brazilian case. Chapter 4 describes the construction of the controversy over GMOs in Brazil. Identifying the relevant actors involved, the story of the campaign is told by taking into account the entry of each collective actor, analysing how they framed the inclusion of GM crops in their struggles and what resources they brought to the campaign. Although the political and economic context of neoliberal adjustments in Brazil was highly favourable to the arrival of the new seeds, social mobilization found a fertile ground among civil society in the 1990s due to the existence of organized movements among family farmers, peasants and agroecology activists. Another three main reasons stimulated the dispute over agrobiotechnology: divergence of opinions among the scientific community, the consolidation of environmentalism in Brazil, and the fact that civil society was keen to explore the new legal opportunities and new institutional channels opened with democratization. They began a national controversy over GM crops, bringing about a judicial action that culminated in a legal moratorium on GM soy. However, there were also precursors, spearheaded by the peasant movement, that politicized the technology in the south of Brazil on the borders to Argentina where GM soy was first cropped illegally. In 1999, the national campaign GM-Free Brazil was officially launched. All this mobilization on the side of civil society was met with an organized reaction on the part of biotech firms, state authorities and farmers who wanted the technology in Brazilian fields.

Chapter 5 describes how Brazil, a case of initial success of social contestation, was converted into a top producer of GM crops. This begins with crucial moments of disputes in the executive and legislative arenas shaping the legal framework, and, ultimately, the governmental decision concerning GM soy: Would Brazil opt for being GM-free or would it officially become a GM producer? Reality and legality did not correspond, with the fight occurring between sanctioning illegality versus transforming it into legality. With the entry into force of a new law in 2005 that officially approved GM soy and the approval of GM corn in 2008, the ambiguities and doubts that surrounded the governmental position on GMOs became clear. Brazil quickly reached the pioneers by adopting the technology. The chapter tells the story of the political struggles behind such abrupt transformation, showing how the dominance of GMOs in Brazilian fields remained contested.

Chapter 6 presents the main findings of the research by means of a systematic comparison, guided by the research questions. It describes the different outcomes and the paths that led to them, and presents the main explanatory factors that account for such differences and changes in trajectories. It addresses the questions regarding the role of the context and the economic structures affecting the politics over GMOs. Finally, it summarizes the findings and suggests an explanatory model to understand the conditions that promote or prevent social participation in shaping agrarian futures. Chapter 7, the final and concluding chapter, elaborates upon the empirical and theoretical contributions made, as well as poses the question of how the research added to an understanding of the more general research problem regarding the democratic possibilities in times of contemporary capitalism.

Notes

1 The use of genetic engineering techniques, characteristic of modern biotechnology, has grown exponentially in the areas of medicine and pharmaceuticals (red biotechnology), industrial processes (white biotechnology) and agriculture (green biotechnology). Biotechnology applied to plant breeding, or agrobiotechnology, means the use of a genetic engineering technique, known as recombinant DNA, to design a seed with specific properties. With modern techniques of gene splicing, a strain of DNA (from another organism) is inserted into a seed so that it will have the desired property expressed by the gene. This is why they are called transgenic seeds. They are also commonly referred to as GM crops or genetically modified organisms (GMOs). The widest applications to date are the property of resistance to herbicide and the production of insecticides by the plant. Here I use these terms interchangeably.

2 'The rural poor is understood here as a highly heterogeneous social category, and they include the peasantry with its various strata, landless rural labourers, migrant workers, forest dwellers, subsistence fishers, indigenous peoples, and pastoralists' (Borras 2009, 9). Such heterogeneity and the growing proximity to urban centres by communication technologies and migration makes the question of a peasant identity a complex issue. Writing on Central America, Edelman states: 'The first thing to acknowledge is that the campesino of today is usually not the campesino of even 15 years ago' (Edelman 2008, 251). Interestingly, the contemporary peasant movements are going through a strong process of identity construction. The movements from Via Campesina in some countries highlight their differences to small farmers and family farmers as they oppose public policies that lead them to pursue the same model of agribusiness but on a small scale (Figurelli 2013).

Chapter 1
Theorizing and researching disputes over GMOs

This chapter presents the theoretical and methodological considerations that guided this study. It is divided into two parts. The first is theoretical and begins with an overview of social research on GMOs. It identifies and describes five prominent analytical approaches used to explain public controversies over GMOs: the political economy of food and agriculture, social studies on science, democracy theory, research on corporations and hegemony, and social movement and peasant studies. The overview concludes with the relevant findings to be considered in a study on GMOs. These are incorporated in a theoretical argument, as well as in the analytical framework. The second part presents the methodological decisions on how to approach the research problem, including the research design, case selection, the time periods for comparison and data sources.

Social disputes over GMOs: an overview and a theoretical road for inquiry[1]

GM crops have been on the market for 20 years with no consensus being reached on their appeal. In fact, dissent still reigns in expert arenas (Bardocz *et al.* 2012; Séralini *et al.* 2014), among trade partners (Palmer and Emmott 2012), and among the public in general (The Economist 2010). This literature review identifies the main streams of social theorizing that have problematized why GM crops are a disputed public topic and a relevant issue for social research. The purpose of the review is to discuss existing scholarship in an organized way, according to broader lines of social theorizing.

Five approaches to explain social disputes over GM crops

The review is structured in five sections. It begins with the macro-sociological perspectives that contextualize the disputes over biotechnology in wider processes of transformations in the political economy of food and agriculture. It then moves onto the institutional culture of science-based policy-making. The next section examines the structural conditions for democratic public debate and participation. The last two sections present explanations based on agency by identifying concrete actors as well as their symbolic constructions and strategies. One strain of research looks at the role of corporations in building networks with experts and policy-makers to promote GMOs as a hegemonic technology. The other stream focuses on the other side of the dispute: the social mobilization against GM crops.

The political economy and political ecology of food and agriculture

In dialogue with rural sociology and sociology of food and agriculture, scholars in the tradition of political economy place biotechnology as a constitutive element of the agrarian basis of world economy in its contemporary formation, which has been in the making since the late twentieth century. While some list GM crops among a number of factors constituting the current agrarian formation, others single it out as a central technology of capitalist agriculture. In both cases, these authors adhere to structural explanations, often reliant on historical accounts of the constitution of such macro-phenomena, coining concepts such as world food system, food regime (Friedmann and McMichael 1989), agrifood system (Magdoff *et al.* 2000) and global commodity chains (Hopkins and Wallerstein 1986). The unit of analysis is focused on global patterns of food production and commercialization, characterized by economic dependence and power asymmetries among countries, which in turn influence each country's ability to shape such patterns, or, conversely, to adapt to them.

Magdoff *et al.* (2000) explain the paradox of increased food production with the simultaneous persistence of world hunger in the agrifood system as a result of increasingly turning food production into a source of profit. There is an increasing commodification of agriculture through the transformation of farming inputs into market products, now in the form of patented biotechnological products. Nature itself becomes the main capitalist dynamic to increase accumulation. Such explanations challenge the argument from Herring (2007) and Paarlberg (2000), who claim that GM crops are part of the solution to global food security.

Kloppenburg (2004) situates the role of biotechnology to processes of commodification in seeds, contributing to overcoming both technical and juridical barriers for capital penetration in plant breeding. The former relates to biological barriers for turning seeds into a commodity, due to their reproducibility that allows farmers to compete with private breeders. With hybridization, breeders increased seed productivity while making seeds sterile and thus forcing farmers into the seed market. Biotechnology draws on hybridization and opens up new scientific ways to make seeds sterile. Juridical barriers have been removed by legal regimes that extend intellectual property rights over genetic material.

A different way in which biotechnology is related to commodification can be comprehended through the concept of the global commodity chain.[2] Most GM crops on the market are not intended for direct human consumption but rather serve as animal feed or industrial food processing, being part of a complex chain between farming and the consumer's table. The positions of countries in the chain affect how biotechnology is perceived and acted upon by policy-makers and the public.

Also drawing from a world-systemic perspective, McMichael (2009) contends that the contemporary world system is based on a 'corporate food regime', characterized by corporate-driven neoliberal regulations aiming at integrating agriculture into the dynamics of the global market. Pechlaner and Otero (2008) make a clear case for the constitutive role of biotechnology in the 'neoliberal food regime', a label they chose due to the role of neoliberal globalism in creating the appropriate conditions for the diffusion of the technology. Free trade regulations strengthen intellectual property rights and promote a lax regulatory system that limits health and environmental protection to scientifically demonstrated risks in order not to create trade barriers. They further argue that biotechnology exacerbates power inequalities between developed and developing countries. While the former have more capital to invest in research and development, acquire patents and influence international regulations, the latter not only lack those possibilities but also comprise the world's agrarian poor population. At the same time, they argue that agrarian change will not follow one homogeneous pattern, as individual country policies provide enough leeway in implementing international agreements. Rather, the trajectory of the food regime will be shaped by the strength of local resistance to biotechnology, particularly if this is associated with protests against the neoliberal restructuring of agriculture.

Calling for an ecological turn in food regime analysis, Campbell (2009) defends the need to make visible the environmental and health impacts of spatially and socially disembedded food relations between consumption and production. Alongside its political economy, looking at the political ecology of GM crops highlights the asymmetrical distribution of environmental burdens such as loss of biodiversity, pesticide contamination and deforestation, which are concentrated in the production nodes of global commodity chains (Cáceres 2015; Otero 2008; Pengue 2005).

Scientific and policy culture

The neoliberal food regime relies on a transnational science-based regulatory framework that unifies different legal regimes under a common denominator: decisions to approve a GMO must be based on the scientific assessment of its environmental and health risks (Motta 2015), or 'physical risks' (Seifert 2011). This creates legal hurdles for applying precautionary measures, as research on risks hardly publish negative results due to conflicts of interest (Diels *et al.* 2011) and exceptional cases face obstacles to publish (Séralini *et al.* 2014).

This makes biotechnology an exemplary topic for science and technology studies (STS). By tracing biotechnology connections to specific actors, artefacts, institutions and interests, STS scholars make visible the configurations that favour biotechnology development and how these empower some actors while excluding others (van Zwanenberg and Arza 2013). Wynne (2001) explains how dominant scientific and policy culture promotes GMOs by maintaining a constructed divide between scientists and the public in terms of facts and objective knowledge, on the one side, and subjective values and ethics, on the other. Labelled as a 'risk' issue, GMOs are transformed into a scientific matter, amenable to objective assessment, whose risks can be reliably known, managed and controlled. At the same time, policy discourse disregards the unpredictability of possible consequences of the unknown. The pervasive myth of 'real versus perceived risks' has been the recurrent explanation of public concerns about technologies. However, STS scholars argue that the public is concerned with a scientific and policy culture that lack a self-critical capacity to recognize the limits of their claims (Wynne 2001), and take for granted the benefits of GM technology (Mayer and Stirling 2004). To them, governmental failure to address public concerns by solely relying on risk assessments has delegitimized these procedures and encouraged polarization.

Indeed, studies on public controversies on GM crops show that science-based expertise, far from solving policy dilemmas, is itself prone to controversy and value-laden disputes and might be used as a political weapon among contenders. Not only do experts have different policy assumptions – varying between a policy commitment to and a policy rejection of GM crops – but also their risk assessments do not correspond to science-based models, being rather informed by the way in which the problem is framed (Bonneuil and Levidow 2012). Policy on GMOs is predominantly framed in a favourable way that is linked to innovation (Van den Daele 2007). For instance, the principle of substantial equivalence has been used to dismiss scientific scrutiny of possible risks as it implies that GM crops are equivalent to non-GM coun-terparts. However, protests challenged its 'technical' basis, transforming it into the starting point for risk assessments instead of its closure (Levidow *et al.* 2007).

Another explanation for public concerns is to consider them as being of an ethical nature (Levidow and Carr 1997). But the created division, knowledge versus ethics, is misleading because public mistrust may well result from ethical and intellectual judge-ments. Surveys on public opinion, such as the Eurobarometer (Gaskell and Bauer 2001), have consolidated such arbitrary divisions. These studies rely on methodolo-gies that inquire into individual attitudes, ignoring the lessons from risk sociology and anthropology, which highlight that accepting risks is intrinsically a social process (Beck 1986; Douglas 2003; Luhmann 2008).

Different risk cultures are the starting point of the work from Sheila Jasanoff (2005), who understands social disputes on GMOs as involving knowledge and normative claims. This is due to biotechnology's hybrid ontology between science, metaphysics and ethics, generating uncertainty and posing the challenge of collect-ive sensemaking. National variations are explained by consistent and robust political cultures, reflected in regulatory institutions: some classify biotechnology as familiar and manageable, while others consider it as new and risky. In any case, regulations are driven by the promotion of innovation that follows market, scientific and bureaucratic rationalities, and not by principles of democratic politics. In sum, scientific and policy culture is reliant on risk assessments that are part of the explanation of public contro-versies about GMOs rather than a solution to it.

Democracy and public debate

With the scientification of politics and the politicization of science, there is an increasing consensus on the desirability of public accountability, debate and critical deliberation on technological and scientific choices. Relying on concepts like public engagement with science, public debate, deliberative democracy and public participa-tion, scholars have analysed social controversies on GMOs, looking at media debates (Horst 2010; Veltri and Suerdem 2013) and government-sponsored public consulta-tions (Boy *et al.* 2000; Dąbrowska 2007; Goven 2003; Magnan 2006). This stream of theorizing sharply contrasts with research on public opinion, as it rejects its methodo-logical individualism, the assumptions of cognitive gap and cultivation effects from the media (Peters 2005).

A main difference lies in highlighting the eminently political quality of public debates on GM crops, in which battles over public meanings are at stake. The hegem-ony of risk discourse in framing the issue has been disputed by appealing to both 'counter-hegemonic expertise' and rights. Peasant expertise links GMOs to issues of food quality, types of agriculture and cultural homogenization in times of globaliza-tion (Heller 2002). Claiming rights implies that only by framing it as a political prob-lem can the public counteract system imperatives from the state and capital to treat biotech crops as a matter of innovation and risks calculation (Anderson 2004).

Differing from explanations for cross-national variations based on robust political cultures, research on public debates over GMOs has shown how social meanings are more plural and fragmented than media discourse or policy dis-course. This varied research has some common denominators. First, they show the constitutive pluralism of public meanings and therefore that public debate will not lead to consensus. Second, this acknowledgement does not invalidate its role in legitimizing democracy. Authors converge in the view that a wide public debate on GMOs should not be restricted to technocrats and experts, which is also desired by the public in their case studies. Therefore, the assessment that such initiatives fail to diminish the controversial character of the issue (Gaskell 2004) misses the point that such plurality and the lack of trust in regulators and in the food industry will not be overcome with participatory exercises alone. Finally, they call attention to the conditions in which consultations are held, namely under urgent political and economic pressures promoting the technology, and to the need to ensure possibilities of rational-critical debate on a regular basis. These structural conditions of contemporary capitalism call into question the possibilities of such debates to influence political decisions. Among the many obstacles for a democratic public debate are the strategies of corporations to align political, material and media interests in favour of GMOs. To these we now turn.

Bio-hegemony: corporations, policy-makers, agrarian elites and experts

There is a growing amount of research on how corporate actors have constructed a coalition with scientists and public officials in order to build consensus in favour of biotechnology. Here, the focus is on the social actors, networks and coalitions that promote GM crops: their strategies, frames and resources.

Glover (2010) studied Monsanto, the leading firm in biotechnology. He argues that despite the aspiration of having a lead role in developing solutions for food security and environmental sustainability, the actual technology development was primarily shaped by technical, commercial and financial considerations. These meant more a continuity of the company's agrochemical profile than a technological rupture in the purported direction. The backlash in Europe, which closed its markets to the technology, forced the company to focus on developing markets. However, this was mainly a transformation in framing, not in actual technological development.

In order to explain how the pro-poor and sustainability framing came to be so accepted by many policy-makers, journalists, scientists and influential commentators, Newell (2009) coined the concept of bio-hegemony. Inquiring how the interests of a dominant sector that mostly benefited from the spread of GMOs became identified with the common good for society in general, he argues that this is achieved through an 'alignment of material, institutional, and discursive power in a way which sustains a coalition of forces which benefit from the prevailing model of agricultural development' (Newell 2009, 38).

In Argentina, the country identified by Newel (2009) as an expression of bio-hegemony, Gras and Hernández (2008) studied how the conversion to GM crops was accompanied by transformations of farmers' identities into that of entrepreneurs together with the public meaning associated with agriculture. Key to the construction of a good image of agrobiotechnology was emphasizing knowledge as a factor of production as well as an ideological norm (Hernández 2007). The new agrarian entrepreneurs resignified their activity as part of a knowledge society, in which farming depends on new management and networking skills. Their identity as modernizers contrasts with that of landholders, in a strategic move to compete not only for rents but for power and legitimacy.

Studying Uganda through the lens of bio-hegemony, Schnurr (2013) explains how the choice of GM technologies was not oriented to solve the problems faced by Ugandan farmers but by a donor-determined (commercial) agenda. This gap between technological promises and farmers' realities resulted from a combination of strategies. By materially co-opting local researchers, firms transformed them into 'organic intellectuals', who used their credibility to promote GMOs. Biotech firms influenced regulations by establishing a fast-track approval of GMOs. They defended regional harmonization in Africa to foster free trade. In the institutional dimension, corporate actors allied with public officials, politicians and state bodies through research funding, cooperation agreements, funding of trips and organization of informal meetings in which social ties and networks are formed. Finally, firms built their discursive hegemony with communication efforts such as special magazines delivered to politicians and to editors and journalists at major print, radio and television companies as well as media campaigns on the radio and television programmes that reached poor farmers.

In sum, bio-hegemony in practice involves the construction of many social relations that secure its material, institutional and discursive power. Far from being a privilege of developing countries, bio-hegemony is played out in various settings. The food regime based on biotechnology is not, however, hegemonic yet, at least not everywhere. It is in the making and it is under dispute, as the literature on social mobilizations against GMOs shows.

Social movements and peasants

Concerned with the role of agency to explain change in the agrifood system, Schurman and Munro (2010) explain how a small group of critics became a global social movement that negatively impacted the plans and fortunes of the biotech industry. Calling attention to the role of ideas, they describe how activists developed a critical interpretation of the technology and established their collective identity as a social movement. The movement began as a highly professional group of scientists and lawyers, relying on strategies of counter-expertise, lobbying and legal action. Among the set of structural factors, they contend that global commodity chains created new pressure points for activists, who targeted the consumer end, namely supermarkets and retailers. Schurman and Munro conclude that the most important impact of activism was to construct the categorical difference between GMOs and non-GMOs, and to affect the industry's plans by making GMOs a politicized and contested technology.

Whereas the historical origins of the global anti-GMO movement are found among the white middle class with highly professional backgrounds in science and law, as well as good institutional positions in the NGO field, the anti-GMO movement only gained an impetus of a mass movement when the issue of GM crops was incorporated into the agenda of movements from the rural poor (Borras *et al.* 2008). Indeed, another perspective on social mobilization against GM crops is found in the work of scholars in the area of rural sociology, agrarian change, peasant studies and development studies.

The class character of mobilization against GMOs was underlined by Scoones (2008) in his study of Brazil, India and South Africa. Large producers of GM crops, these countries are characterized by rural poverty and a dual rural structure of agribusiness, on the one side, and small and subsistence farming, on the other. Each side of the controversy, aligned with specific agrarian interests and politics, claimed to represent the material interests of smallholders. Scoones poses the question of whether the transnational, elite character of activists against GMOs affects their legitimacy to speak for small farmers. To him, the anti-GMO movement represents a new type of legitimate political expression, built on transnational solidarities and shared goals. It demands the opening of the debate in a neoliberal context that pushes to close it. In particular, it challenges the restriction of the debate to knowledge issues based on mainstream science, and links it to politics, ethics and values, and rights and justice.

Indeed, agrarian movements have extended the debate on the technology to include intellectual property rights, corporate control, the rules of global trade of agrarian products, as well as the future of small-scale agriculture and of agrarian society. To them, GMOs are but one element of the agrifood system against which they fight, one item in their demands for an alternative grass-roots agrarian model, subsumed under the frame 'food sovereignty'. Martínez-Torres and Rosset (2010) argue that Via Campesina, the most prominent of the agrarian movements, makes more use of confrontational strategies, such as radical mass demonstrations, rather than taking the insider role of lobbyists, although they also engage in the latter; when doing so, they focus on political and moral arguments and avoid a technical frame of the debate.

Critical perspectives on food sovereignty, however, warn that by rejecting biotechnology *tout court*, social movements and scholars run the danger of repeating the same mistakes of modernization theories, inverting the value ascribed to tradition and modernity (Bernstein 2014). Social movements are aware of such a danger. Heller (2013) analyses the engagement of the French agrarian union Confédération Paysanne in technical and political debates about GMOs. Focusing on risk issues, union leaders relied on an instrumental rationality, with lobbying strategy to influence policy bodies. Simultaneously, the direct action wing demanded a general ban on GMOs, based on a solidarity-rationality concerned with cultural and economic effects of the new technology over land, work and quality of life in a post-industrial agricultural condition. They shifted the issue to a domain of peasant expertise, challenged regulatory boundaries between economic and ethical arguments, and called attention to the effects on livelihoods of peasants and indigenous communities in the Global South. The solidarity rationality has increasingly informed concerns about risks, substituting an individualistic risk logic for a humanist perspective on effects on public health and environment. Kinchy (2012) reaches similar conclusions, arguing that environmental concerns are intertwined with a critique of power and inequality in the world agrifood system. He researched how social movements engaged in strategies of counter-expertise, looking for openings to construct alternatives to sustainable, non-chemical, economically viable rural livelihoods. These works show that peasant movements are not anti-science; rather, they dispute the dominant discourse and policy on GMOs that gives primacy to a specific type of scientific expertise.

Instead of focusing on activists, an alternative approach has been to inquire directly into the livelihood strategies of smallholders in the countryside. Fitting (2011) argues that the future of the biodiversity of maize depends more on the practices of Mexican peasants than on regulations on GM imports. The problem is what she calls the disjuncture between the debates on GM corn taking place in the urban areas, newspapers, and political and academic arena, and what happens in the countryside. In other words, prohibiting GM crops in Mexico will not suffice in alleviating the struggle of Mexican peasants to maintain their livelihoods in the context of a neoliberal corn regime. In a similar vein, Lapegna (2014) makes a plea for a global ethnography as a method to investigate GMOs not only as global entities but also as cultural objects and local practices, in particular political and social contexts. He contends that among the limitations of macro-views is that they have little to say about how global processes are differently experienced locally, including variations between resistance and accommodation among subordinate actors.

This body of research converges in the finding that consumers and environmental NGOs have widened the debate, as well as achieved outcomes such as the incorporation of the precautionary principle in some regulatory regimes. By contrast, peasant movements mobilized to shift the debate from a technological frame to a political debate on the exclusionary effects of the agrarian policy have hardly succeeded in terms of policy change (Newell 2008). Nevertheless, they keep mobilized for their goal of shaping an alternative food and agrarian model, with varied degrees of success in different countries. Their processes of organizational building, network formation, identity construction, meaning-making and mobilization can also be considered outcomes, in the sense that they contribute to institutional and cultural changes in the wider social context (Giugni 1998). Indeed, the mere existence of peasants, once deemed to disappear, and their increasing social mobilization challenge not only the proponents of GM crops but also social theory in general (McMichael 2008).

Some conclusions from previous research

As this literature overview shows, the social scientific study of GMOs offers various lines of inquiry. Each of these strands of research has contributed to understanding the emergence and dynamics of social struggles on GM crops. A first concluding remark is that even though each scholar presents one analytical focus based on her or his theoretical affiliations, most of those doing research on GMOs tend to point to the other factors as playing an important explanatory role.

Authors applying structural explanations of political economy also highlight the role of science in legitimizing the current food regime, while, at the same time, arriving at the conclusion that structures can be challenged and have their trajectories changed. It would be mostly likely for this to happen when there is an alliance between struggles against biotechnology and those against neoliberal restructuring of agriculture. Scholars studying the social construction of a policy culture based on scientific discourse recall that dominant institutions aiming at legitimizing biotechnology cannot evade public concerns and demands for accountability and participation. All agree that social actors, with unequal capacities and power, shape these macro-phenomena and that they result from social disputes: on the one side, corporate lobbying, experts and policy-makers that promote it; and, on the other, resistance to it by social movements. Along the same line, those studying corporate strategies are aware of the structural asymmetric conditions among countries in their autonomy and power to act upon corporate pressure. Social movement scholars emphasize structural conditions for social mobilization, with the caveat that structures must be interpreted and mobilized by social movements as constraining or facilitating their strategies (Gamson and Meyer 1996). Among these strategies, they have targeted global capital and mobilized scientific expertise to challenge the dominant scientific and policy culture. In sum, there is a growing consensus on what are the relevant factors and on a dialectic tension between them.

A second conclusion regards the multiple levels of analysis. One cannot start to understand biotechnology from a sociological perspective without considering the global dimension of the material and cultural support that has been promoting the technology in the last three decades. On the one hand, global developments in the agrarian economy include the commodification of seeds, from which biotechnology is the last trend. International legal instruments construct a policy culture based on science and risk as criteria for decision-making. On the other hand, it is not sufficient to remain at a global level of analysis, as national states play a key role in such international developments. This means that there is some – asymmetrically distributed – room for autonomy in the form of individual states negotiating and interpreting such global trends when deciding which role biotechnology should take in their development strategies.

This brings us to the national level of analysis and to the embeddedness of markets, science and politics into specific contexts with different trajectories. Countries are differently positioned in the agrarian global market; they also have different democratic cultures and practices. National processes of public deliberation and social mobilization have challenged the legitimacy of global market dynamics and scientific discourse as determinants for national policies on GMOs, with varied results. At the same time, the social and environmental effects of the global expansion of biotechnology are mostly concentrated at the local level, in the rural and suburban communities that surround GM fields. This level has given rise both to the emergence of grass-roots movements against GMOs and to the adaption of communities to the effects of the corporate food regime. Finally, for all these reasons, it is necessary not to treat national case studies as self-contained units of analysis by looking at how events in one place affect others and also by situating countries in an asymmetrical world agrifood system.

A third conclusion from the reviewed scholarship is that social disputes over GMOs will remain controversial as there is neither a sole explanation nor a single solution for them: these are ultimately political assessments. Controversies vary according to country positions in the world agrifood system and to political cultures on how to make sense of the new technology. Within national borders, research has converged in the finding that neither expertise nor participatory exercises have contributed to decreasing the controversy. Among the reasons are the fragmentation and plurality of publics as well as distrust in scientific and policy culture that do not recognize the limits of their claims while promoting GMOs. Whereas biotech corporations and their networks attempt to close the debate and construct a hegemonic public meaning for GMOs that reflects their material interests, social movements struggle to enlarge the scope of the debate beyond regulatory boundaries. Movements and scholars call for the need for the democratic definition of technological development and agrarian futures. In sum, it seems that GMOs will remain a topic of public and academic attention in the future, as divergent opinions on agrarian models dispute the solution for pressing issues of the global agenda, such as food security and climate change.

A theoretical road for inquiry

Based on the lessons learned from previous research, this book combines a macro-sociological analysis – both material and symbolic, namely of political economy and ecology, scientific culture and democratic conditions – with an actor-centred analysis on the meso- and micro-sociological levels. These analytical

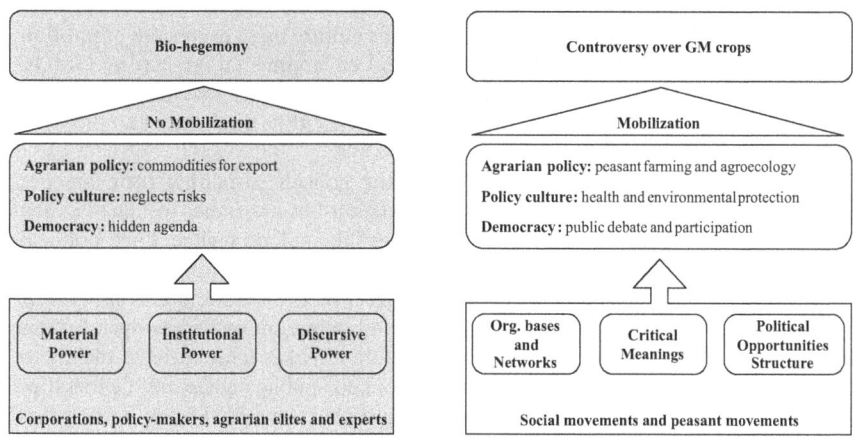

Figure 1.1   Paths and mechanisms leading to the outcomes *bio-hegemony* and *controversy over GM crops*

Source: Own work.

factors are interpreted according to their contribution to different paths and out-comes in the social disputes over GM crops. On the one end of the spectrum, there is a construction of a hegemonic, corporate-controlled, neoliberal food regime based on agrobiotechnology, in short, a biotech food regime[3] in a position of bio-hegemony. On the other end, there is the construction of GMOs as a political problem, subject of public debate, democratic decision-making and target of social mobilization, defined as controversy over GMOs. In an ideal-typical fashion, each of these outcomes would correspond to a particular expression of the analytical factors (Figure 1.1).

Bio-hegemony implies that the political economy of food and agriculture in a given country is characterized by the state promoting the widespread adoption of biotech crops in commodity production, high presence of transnational corpora-tions, large-scale farming, incorporation of neoliberal international free trade reg-ulations, and strong integration of agrarian markets in global commodity chains. Socio-environmental impacts – such as land concentration, conflicts over land, and environmental and public health impacts – of these agrarian transformations are 'externalized' and made invisible. By contrast, in a controversial situation over GMOs, alternative agrarian models compete for public attention and policy sup-port. The state promotes agrarian policies aimed at small-scale and family farming, directed towards the internal food market, including the support for agroecology, the use of non-commercial seeds, and special policies for land reform and protection of land rights.

In a hegemonic biotech food regime, a policy culture neglects the existence of risks associated with biotechnology and scientists act as promoters of GMOs. This pol-icy culture is based on a distinction between experts and lay public while deflecting any critiques of their monopoly of knowledge as well as rejecting demands for public accountability and participation. Therefore, it functions as a mechanism for eschewing public debate by framing GMOs as a technocratic issue. Conversely, in a controversial situation over GMOs there is room for a scientific culture that ensures the promotion of continuous research for dismissing health and environmental impacts of GMOs. Scientists feel the burden of public accountability and of their role in a democracy. Controversy implies, at least, that there is room for scientific dissidence and plurality of views.

In bio-hegemonic situations, GMOs and the policies sustaining them are not open for public debate and democratic participation but are rather insulated in the hands of state bodies committed to the biotech neoliberal food regime. Their political decisions appear as an inexorable means of agrarian modernization and innovation. Controversy exists when those policies are objects of democratic decision-making, in which agrarian development and technology adoption are treated as open futures dependent on present political decisions. Therefore it makes sense for civil society to take part. This assumes that structural conditions for public debate and deliberative democracy are given, that is to say, that such processes are guarded from urgent political and economic pressures for approving technology and from the influence of political and material power. In this case, the resulting public decisions and opinion acquire democratic legitimacy.

Networks of corporations, policy-makers, agrarian elites and experts are the main agency behind the constitution of bio-hegemony. They deploy material, institutional and discursive power in order not only to ensure that the more resourceful will see their positions prevail but also to build a consensus that their interests in promoting GMOs also reflect society's general interests. They may succeed in building a symbolic order that provides legitimation to their material dominance, but this is not democratic legitimation. To the contrary, these actors will avoid democratic processes and transparency, and make all efforts to preclude the construction of a controversy over GMOs. Reactions to social mobilization might include attempts to suppress or silence disputes through strategies of increasing material, institutional and discursive power to win the support from recalcitrant politicians in the legislative and to silence oppositional governmental bodies and dissident scientists. Strategies targeted at civil society include the criminalization of social movements, violence against activists and silence of the media on their claims and protests.

Social movements and the agrarian poor are the main agency to contest the agrarian transformations leading to bio-hegemony and to bring the issue of GMOs to debate. The defining mechanism behind constructing and sustaining a controversy is social mobilization, in particular when this translates into the launching of a campaign, the building of a coalition among different social movements and organizations, the brokerage between urban and rural sites, and the construction of collective identities and political subjects to claim rights. Their social mobilization depends on organizational bases and networks, critical meanings and political opportunity structures. These are ideal types of outcomes and paths. The same actors that create a controversy can thus undermine bio-hegemony just as the actors supporting the latter might act to demobilize social movements and close a controversy, as depicted in Figure 1.2.

Finally, the political economy of food and agriculture also influences the context for social mobilization. It is thus necessary to understand the structure of political opportunities as being embedded in the structures of global capitalism in order to identify how it contributes to an outcome of bio-hegemony or controversy over

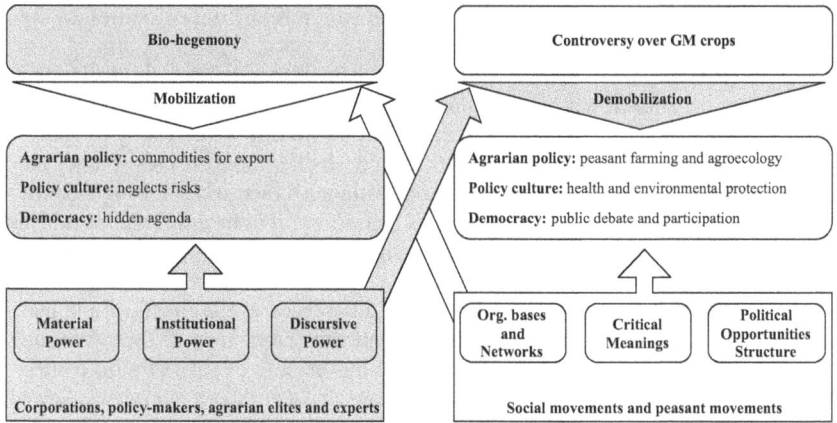

Figure 1.2   Changes in mechanisms leading to the outcomes *bio-hegemony* and *controversy over GM crops*

Source: Own work.

GMOs. Inquiring into this relationship has four theoretical implications. First, countries cannot be compared equally without taking into account their asymmetrical position in world agrarian markets. Second, given the connections between commodity production and consumption, outcomes of social movements' campaigns in one country might affect the GM policy and social mobilization in another country. Third, the concentration of socio-environmental effects at the production nodes of the chain implies that the local level of political action is especially relevant for contesting bio-hegemony or accommodating it. However, the fourth argument is that long-lasting solutions to local problems might be conditioned on decisions taken nationally by foreign countries or at the global level. This point brings us back to the first statement: political opportunity structures in countries specialized in the production node of the global commodity chains present additional constraints for movements succeeding in their demands.

Such a complex theoretical approach incorporates lessons from past research into a coherent argument that puts into relation all explanatory factors, including their specific contribution to the construction of GMOs as a political problem or as being in the best interest of general society. As the main mechanism explaining the differences in outcomes is mobilization – and the lack thereof, or even, demobilization – the relevant factors that account for such processes will be described below and organized in an analytical framework.

Analytical framework

Research on social mobilization against GMOs can be classified according to the following categories: (1) who are the activists (their social and organizational bases, and their collective identities); (2) how do movements act to contest GMOs (their strategies or action repertoires); (3) what meanings do they associate with GMOs (the issues and frames); (4) what are the outcomes; and (5) what structures of opportunities do they have and what threats do they face.

The first category addresses who are the activists. The study identifies their social networks and organizational bases by looking at the grass-roots bases, professionalization and the alliances between urban and agrarian sites. It is interesting to analyse whether activists construct collective identities and whether these are class-based, gendered or cultural. Among the rural poor, some distinguish themselves through identity building in regards to their symbolic work as peasants and indigenous peasants. This collective identity is built in connection to their lands and how they relate culturally to them (Escobar 2001).

Among the rural poor, there are also gender differences, as the effects of and reactions to agrarian transformations are often gendered (Doss *et al.* 2014). Although women have been very active in agrarian social movements, scholarship on contestation over GMOs still lacks a gender perspective. My fieldwork and interviews showed that one of the strongest mobilizations occurred when women brought to the public sphere the threat they felt as mothers, care workers and those traditionally responsible for the well-being of the family; they claimed that agribusiness was severely affecting the health of family members because the numbers of abortions and malformations increased in areas surrounded by GM soy fields. The gendered identity of mothers became mobilized as a political identity to fight the agrarian model. As a common feature of other women's movements (Jelin 1990), their identity as mothers involved an ethical dimension that appealed to fundamental values (the right to a dignified life) as well as a demand not for political power but for autonomy in conducting their lives, that is to say, the political and economic systems should not interfere with it.

A second category relates to how do movements act to contest GMOs, including their strategies or action repertoires: from the use of counter-expertise (Arancibia 2013; Kinchy 2010), more institutional means such as lobbying and legal action (Bissoli 2013), and alliances with parties to mass mobilization and direct action, including the use of disruptive means and violence. Among the rural poor, scholars distinguish between 'everyday peasant politics' and 'advocacy politics' (Kerkvliet 2009), the latter being organized collectively, taking the form of social movement, and the former being best characterized as 'resistance', in the sense of James Scott's (1987) 'weapons of the weak'. Although not consciously intended as a strategic action, as often assumed by social movement research, the agency of thousands of peasants who resist forced evictions, who maintain seed saving and exchanging practices, and who practise another type of agriculture, represents a counter-hegemonic agency that resists the expansion of the corporate food regime. In researching the conditions for building hegemony, it is also relevant to identify when peasants accommodate to the dominant food regime.

A third category concerns the meanings activists associate with GMOs, what issues they raise, how they frame it and what are the framing activities. In a context of neoliberal regulations that bring policy decision and debate to a close, thereby fostering an expert policy culture, it is interesting to analyse how activists engage in techno-scientific debates, and, moreover, whether they challenge the restriction of the debate to risk issues based on mainstream science by linking it to politics, ethics and values, and rights and justice.

Frame analysis will draw on the extensive research and categories developed by Snow and colleagues (Benford and Snow 2000; Snow and Benford 1988, 1992; Snow *et al.* 1986) such as core framing tasks (diagnosis, prognosis and motivation); frame resonance or potency due to credibility (frame consistency, empirical credibility, credibility of frame articulators) and salience to targets (experiential commensurability, centrality, narrative fidelity); and frame alignment processes (frame bridging, frame amplification, frame extension, frame transformation). It also draws on the work of Gamson (1992) concerning framing and collective action frames, specifically in the identification of the three elements of collective action frames: the injustice component, adversarial component and agency component.

The fourth category is about the types of outcomes movements achieve: if the technology became a topic with public attention and contested meanings, reaching public and political agendas; whether activists influenced policy decisions; and if they altered the behaviour of economic actors regarding GMOs. However, in a more general way, it is relevant to identify whether activists managed to challenge the biotech food regime through their mobilization, as stated before, since the mere existence of contention can undermine the strategies for building consensus and hegemony over GMOs.

The fifth category is the structure of opportunities and threats, including the reactions of authorities and opponents. The concept of political opportunity structure refers to the factors responsible for the prospects of success of social mobilization in achieving its desired goals, which are (1) opening of access for the participation of new actors; (2) evidence of political realignment within the polity; (3) availability of influential allies; and (4) emerging splits within elites (Tarrow 2011). There are new concepts that expand the original concept, particularly 'discursive opportunities' (Bröer and Duyvendak 2009; Ferree *et al.* 2002; Koopmans 2005) and 'legal opportunities' (McCann 2006). Opportunities might also be created by mobilization, which in turn might form cycles of contention when campaigns and coalitions are formed, action repertoires and meanings spread, and opportunities for mobilization are more generalized.

Threat is defined by the risks or costs of mobilization or its absence (Tarrow 2011). One important source of threats is the reactions of authorities and opponents. These can act to control or suppress dissidence in a direct way through outright coercion, or in a more indirect way to discourage actions by increasing costs of organization and mobilization, for instance, by creating financial, administrative and legal hurdles. A main proposition for studying contentious politics (McAdam *et al.* 2001; Tilly and Tarrow 2007) is to use these concepts in a relational way by identifying mechanisms of mobilization and demobilization. The former includes campaign building and coalition formation and the latter might occur when there is repression, control of dissidence, facilitation (when some claims are satisfied and contenders retreat), exhaustion, radicalization and institutionalization (Tarrow 2011). The analytical categories and subcategories for social mobilization are summarized in Table 1.1.

The described factors that affect the structure of opportunities and threats restrict the context for movements' actions in the (national) political process; however, researchers have long emphasized that the structure of opportunities for social movement action is not solely determined by national factors but also by political, economic and cultural processes taking place in the global context (Della Porta *et al.* 1999). Common topics for analysis have been the articulations between transnational, national and local activism and the diffusion and co-construction of frames, strategies and actions repertoires. This study will analyse national political processes embedded in world agrifood systems by identifying the political economy of countries according

Table 1.1   Analytical framework to explain social mobilization over GMOs

Categories	Subcategories
Social networks and organizational bases	Grass-roots bases
	Professionalization in NGOs
	Alliances urban-rural
	Class character
	Gender
Action repertoires	Use of disruptive direct action
	Lobbying
	Legal mobilization
	Mobilization of expertise
	Resistance or accommodation
Meanings	Core framing tasks
	Frame resonance or potency
	Frame alignment processes
Outcomes	Issue attention and agenda setting
	Policy change
	Behaviour change among economic actors
	Social mobilization
Structure of opportunities and threats	Access to decision-making arenas
	Political realignments
	Influential allies
	Splits within elites
	Discursive opportunities
	Legal opportunities
	Existence of campaigns and coalitions
	Repression and control, facilitation, exhaustion, radicalization and institutionalization
Political economy of food and agriculture	Influence of power from commodities exporters in national political process
	Position of countries in global commodity chains

Source: Own work.

to their positions in global commodities chains while also considering impacts at the subnational and local levels.

Research design

This research draws on traditions of methods that conduct macro-analytical qualitative comparisons (Berg-Schlosser and De Meur 2009; Mill 1843; Przeworski and Teune 1970; Rihoux and Ragin 2009; Skocpol 1976; Skocpol and Somers 1980), by adopting a research design of Most Similar Different Outcomes (MSDO). This choice has the following implications. First, it is a case-oriented approach (rather than a variable-oriented) since it aims at understanding and explaining only the cases considered. This requires in-depth knowledge of each case, which are considered complex and unique units. Second, it entails a specific method, namely the comparison of their similarities and differences. However, the research is not limited to the description of cases. Thus the third implication is that the method entails an analytical procedure due to its explanatory purpose: it aims at identifying instances that are relevant to explain the variation in the phenomena. Such identification contains both deductive and inductive steps. There is deduction in the selection of conditions, which is guided by theory. But the process of contrasting the cases is grounded in case knowledge and involves going back and forth between theory and empirical evidence. Theory serves as a good starting point to open up the analytical framework in search of relevant conditions. The process of reaching explanations for the differences in outcomes is rather inductive.

This method offers the tools to tell these stories in an analytical way, thereby looking for explanations. Still, these general statements are valid for the limited conditions of space and time, which means that the research results only have internal validity. However, they might guide further research that expands the universe of cases. This comparative research design is adopted with the caveat that it is not restricted to the search for endogenous explanations for each case, as other processes take place across and beyond national borders, addressed in the multiple levels of analysis.

Case selection

The choice of countries was theoretically oriented in order to serve as contrasting instances of the object of study. Considering the leading producers of GM crops (see Table 1.2)[4] as the universe of possible cases, comparison was a reasonable choice to approach the research problem. The next step was to choose two cases that varied in the construction of GMOs as a political and controversial problem, as a positive outcome, and in its hegemonic adoption, as a negative outcome. The defining mechanism that might lead to controversy is social mobilization. Among the countries listed, Argentina and Brazil offer the best combination, as they are relatively the most similar ones in many respects, including in terms of positions as GM producers and the biotech crops authorized for cropping, while they significantly diverged in the trajectory leading to this position.

Brazil is a case of controversy, at least in its initial trajectory, due to the existence of social mobilization over GM crops,[5] where there were political claims, protests and divergence among experts that threatened the interests of the pro-GMO coalition. There was clear dissent on the issue, with varied types of actors having a say, and media coverage of GMOs as a political problem. Accordingly, actors in this dispute regarded the politics and policy for GM crops as open and in the making. This is evidenced in the various political decisions, parliamentary debates, law-making and judicial decisions. Argentina shows a contrasting outcome of bio-hegemony,[6] where GMOs appear as uncontroversial, one-sided, with the dominant public meaning reflecting the interests of agrobiotechnological firms, agribusiness farmers, experts and policy-makers. The few protests and contentious claims were not visible in the public sphere, where the pro-biotech policy appeared unchallenged, scientific certainty on

Table 1.2   **Global areas of biotech crops in 2014 by country (million hectares)**

Position	Country	Area (million hectares)	Biotech crops
1	United States	73.1	Maize, soybean, cotton, canola, squash, papaya, alfalfa, sugar beet
2	Brazil	42.2	Soybean, maize, cotton
3	Argentina	24.3	Soybean, maize, cotton
4	India	11.6	Cotton
5	Canada	11.6	Canola, maize, soybean, sugar beet
6	China	3.9	Cotton, papaya, poplar, tomato, sweet pepper
7	Paraguay	3.9	Soybean, maize, cotton
8	Pakistan	2.9	Cotton
9	South Africa	2.7	Maize, soybean, cotton
10	Uruguay	1.6	Soybean, maize

Source: Adapted from James (2015).

benefits seemed to reign, and risks were considered nonexistent. The position from pro-GMO actors and their discourse on the economic benefits of GM crops prevailed, including the tropes of innovation, competitiveness and knowledge economy.

These different trajectories are reflected, to some extent, in the rate of adoption of GM crops. Argentina is part of the group of the six 'founder biotech crop countries' (James 2014) that simultaneously adopted GM crops in 1996; until 1995, it was the country in Latin America with the highest numbers of field trials, whereas there were no trials in Brazil (James and Krattiger 1996). The Argentinean adoption was progressive and linear; in 1999, soybeans had crossed the threshold of 70 per cent of production, which happened ten years later in Brazil, in 2009. GM soybeans have dominated Argentina soy production since 2001. According to data from the biotech industry (James 2015), virtually all acreage of soybeans (100 per cent), maize (94 per cent) and cotton (99 per cent) has been converted to the new technology in Argentina. Brazil had a quicker rate of adoption, catching up with Argentina in area planted (Figure 1.3). Today, GM crops occupy a significant acreage of soybeans (88 per cent), maize (75 per cent) and cotton (50 per cent), but there is still room for a market niche of non-GM products.

At the same time, their comparison is facilitated by their similarities vis-à-vis the universe of possible cases for selection, reducing the overdetermination of causes.[7] In the last 20 years, Argentina and Brazil have led among South American countries in the increase in area planted with GM soy, in volume produced, in the area cultivated with the crop, accounting for 40 per cent and 50 per cent, respectively, of soy production in the region in 2010 (Catacora-Vargas *et al.* 2012). The agrarian change that accompanied the expansion of GM crops had similar social and environmental impacts in both countries. The expansion followed two patterns: the substitution of other agricultural activities or the conversion of previously natural habitats into agricultural land (Catacora-Vargas *et al.* 2012). The launching of new varieties of soybeans adapted to new lands both in Argentina and Brazil allowed for the expansion of the agrarian border, causing deforestation, while the intensive use of land led to processes of desertification in the Argentinean Pampas, in the Brazilian *cerrado* as

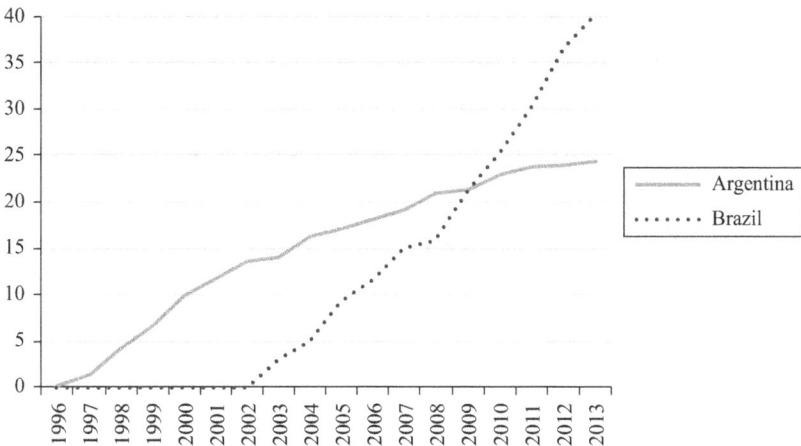

Figure 1.3   Adoption of GM crops in Argentina and Brazil (million hectares)
Source: Own creation based on data from ISAAA.

well as the Amazonian regions. Monocropping increased the ecological debt, as it also exports phosphorous, nitrogen and other nutrients (Pengue 2005).

The use of the technological package transgenic seeds resistant to chemicals led to soil erosion and decrease in soil fertility, increasing the use of fertilizers (Pengue 2005), and to the development of glyphosate-tolerant weeds, to which farmers responded by increasing the dosage. In the last decade, the Brazilian market for pesticides increased 190 per cent vis-à-vis the world market increase of 93 per cent, and since 2008 Brazil has surpassed the USA as the first consumer of pesticides in the world. The increase in the volume of pesticides has been higher than the increase in area planted (Carneiro *et al.* 2012). In both countries, the distribution of pesticide use according to crops reveals that soy and maize are the most chemical-intensive crops, accounting for 40 per cent and 12–15 per cent, respectively, of the total consumed. According to the Chamber of Plant Health and Fertilizers (CASAFE), in Argentina 317 million litres of pesticides were used in 2012, compared to 123 million litres in 1997 (Kleffmann & Partner SRL-Kleffmanngroup 2013). In Brazil, 852.8 million litres were used in 2010 in comparison to 599.5 million litres in 2002 (Carneiro *et al.* 2012).

The increase in area planted does not mean that more farmers are taking part in the production, as large-scale industrialized production is responsible for the majority of the land use. As a consequence, there is a pressure on land prices and changes in land tenure. In Brazil, the structure of landholdings has consistently increased in area and number of properties from 1998 to 2012, but it has not changed the concentrated structure of distribution, and the GINI coefficient for land has remained above 0.8 in the country (Girardi and Vinha 2013). There is no official data in the GINI coefficient for land in Argentina. According to calculations based on census data, from 1998 to 2002 the number of properties decreased by 24.5 per cent while becoming more concentrated: 58 per cent of properties smaller than 100 ha occupied 3 per cent of land, whereas 4 per cent from all properties are bigger than 2,500 ha and occupy 63 per cent of the country's land (REDAF 2014). Sotomayor *et al.* (2011) estimate that the average size of agrarian properties in Argentina is 562 ha, while in Brazil it is 63 ha.

In short, searching for the causes of the emerging social disputes in Brazil that constructed GM crops as a political problem, Argentina serves as a contrast as a case of low or absent mobilization and public debate. Conversely, Brazil provides a contrast to identify explanatory factors for the construction of a hegemonic discourse on GM crops in Argentina. Both outcomes can be explained, that is to say, the causes for the emergence of a controversy over GM crops in Brazil and the causes that prevented it in Argentina and that created bio-hegemony, as well as the changes in each, as Brazil turned into a world GMO producer and social mobilization over the issue increased in Argentina.

Comparison over time

In order to account for such changes, the research design incorporates a time dimension, comparing two phases. The first comprises the period from the mid-1990s to 2002. In both countries, governments followed the directives of the Washington Consensus by adopting structural adjustment policies: macroeconomic stabilization, state reform, privatization and trade liberalization. These reforms also affected agrarian policy. At this time, firms started petitioning for state approvals for the commercial cultivation of GM crops. In Argentina, this context favoured the installation of biotech food regime, generally presented as a successful revolution in agriculture, and social mobilization grew as a reaction to its consequences. In Brazil, previously mobilized social movements deterred the introduction of GM seeds through a judicial moratorium. Therefore, this first phase is called bio-hegemony in Argentina and controversy over GMOs in Brazil.

The second phase starts in 2003. In Argentina, it marks the end of a transitional period. Export taxes on soy were reintroduced in 2002 to improve state finances and, in 2003, after almost two years of economic and institutional crisis, a new coalition government was elected. It is the start of a period of consolidation of the biotech food regime, as its hegemony became associated with its role in the economic stabilization of the country. At the same time, this phase is characterized by the accumulation of grievances and protests related to the agrarian model. However, social mobilization did not succeed in constructing a controversy on GMOs. This is why Argentina remains a case of bio-hegemony, but qualified with an increasing process of mobilization. In Brazil, 2003 was a clear turning point: the newly elected government lifted the moratorium on GM soy. With the entry into force of a new law that officially approved GM soy, in 2005, and with the approval of GM corn in 2008, the ambiguities and doubts that surrounded the governmental position on GMOs were cleared. Brazil joined the group of leading producers of GMOs. Nevertheless, social mobilization disputed every law and every regulation while achieving some concessions, and this is why the new outcome in Brazil can be considered a dominance of GMOs but not yet bio-hegemony (Table 1.3).

The time dimension allows for the comparison of how different contexts in the political economy in these countries affected the social disputes over GMOs. Whereas the first phase is commonly considered the neoliberal decade, the second phase is marked by a promise of change in both countries. The year 2003 heralded the political phenomena that became known as *kirchnerismo*, in Argentina, and *lulismo*, in Brazil,

Table 1.3   Phases of social disputes over GM crops in Argentina and Brazil

	Outcomes	
Context	Argentina	Brazil
1995–2002 Political economy of Washington Consensus	Bio-hegemony *(Chapter 2)*	Controversy over GMOs *(Chapter 4)*
2003–2013 Political economy of Commodities Consensus	Bio-hegemony with mobilization *(Chapter 3)*	Dominance of GMOs *(Chapter 5)*

Source: Own work.

named after the strong personality of each president. There is an academic (and political) debate over the points of continuity and of rupture with previous governments (Gonçalves 2012; Menezes and Palermo 2012; Murillo and Levitsky 2008; Nobre 2013; Sampaio Jr 2012; Singer 2012; Svampa 2008). However, there is some consensus that these governments have pursued policies that changed the political economy (Bresser-Pereira 2012; Wylde 2010). Contrasting the previous Washington Consensus, which advocated the reduction of the role of the state in the economy and in social welfare, the new governments defended state intervention in the economy in order to promote economic growth and expansive social policies to reduce poverty.

The first change included growth promotion based on the extraction of natural resources. In some cases, this was accompanied by advanced control of monopolies of extractive activities, through direct participation or by increasing taxation. State finances started to rely on improving the trade balance, with a competitive exchange trade, and the state actively promoted primary exports.[8] Svampa (2012) called this 'Commodities Consensus'. The boom in commodity prices allowed governments to expand social expenditure, the second change. These governments created an intersection between an agenda of poverty reduction and the extractive model, exploring possible complementarities. Data on commodity dependence and poverty reduction indeed shows a difference over time between the two governments.

Commodity dependence: export, gross domestic product, taxes

Dependency on agrarian commodities (in which biotechnology is applied) is not new in these countries; it grew in the twenty-first century, after a decade in which it had been relatively lower, as measured from participation in gross domestic product (GDP), exports and state revenues. Argentina and Brazil are classified as having high commodity dependence, with commodity exports (in millions of US dollars) accounting for, respectively, 67 per cent and 65 per cent of merchandise exports in 2012 and 2013. Among all commodities, agrarian products destined for feed and food correspond to 81 per cent (Argentina) and 52 per cent (Brazil) (UNCTAD 2015). During the period from 1990 to 2011, soy and corn and their processed products were responsible for almost 25 per cent of Argentinean and 10 per cent of Brazilian export values (Figure 1.4).

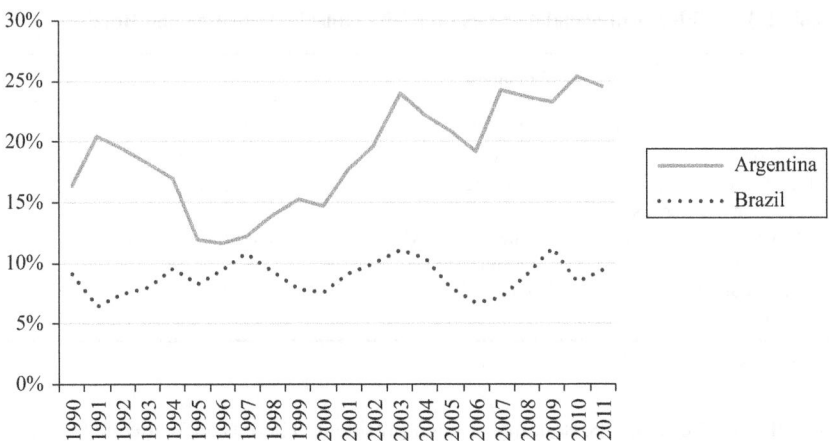

Figure 1.4   Participation of soy and derivatives in exports to the world market (%)
Source: Own creation based on data from FAO.

The participation of soybeans production in the GDP is not so significant as it is in exports. According to data from the International Monetary Fund (IMF), it grew significantly during the years of economic crisis in Argentina, from 1 per cent in 2001 to 5 per cent in 2003, and then dropped, remaining between 3 and 4 per cent. In Brazil, it followed a similar path, but at a much lower level, reaching 2 per cent in 2003 and since then has remained just above 1 per cent. From the Argentinean production, it is estimated that 90 per cent of soybeans is destined for export, from which 78 per cent is in the form of meal or oil (*Veja* 2011). In 2013, 41.6 per cent of Brazilian soybeans production was destined for export as raw material; about 60 per cent was directed to domestic processing as meal and oil, from which 56 per cent and 36 per cent, respectively, were exported (ANEC 2014; Embrapa 2014).

The more its production is destined for foreign markets, the higher the possibility of the Argentinean state to tax it. The export tax rate is currently 35 per cent on average for soy and soy products. Export rates accounted for 7 per cent of state revenues in 1991 before the expansion of soy in Argentina and were suspended during the 1990s to be reintroduced in 2002. From then to 2012, it oscillated between 8.29 per cent and 13.18 per cent of state revenues, dropping to 5.90 per cent in 2013. The participation of soy products in the overall export taxes varied between 40 per cent and 60 per cent between 2003 and 2012, reaching 80 per cent in 2013 (Rauchecker 2015). Brazil does not tax agrarian exports,[9] but has a complex tax system on agrarian products, following the general cascading tax system that taxes one product at different stages of the supply chain, resulting in a cumulative process.[10] This complexity poses an obstacle to finding data on the participation of the soy and corn complexes in tax revenues. One can assume that as soy production increases with the majority destined for internal processing, so does the state revenue from it.

Poverty reduction

The rupture with past governments is attributed to the high impacts of social policies that reduced poverty in these countries. According to the Socio-Economic

Table 1.4   **Poverty indicators in Argentina and Brazil (1995–2010)**

Countries	Years	Population living on less than		Total
		1 dollar per day	2 dollars per day	
Argentina	1995	3.88	7.06	10.94
	2003	9.79	17.92	27.71
	2010	0.92	1.87	2.79
Brazil	1995	11.26	20.81	32.07
	2003	11.21	20.60	31.81
	2009	6.14	10.82	16.96

Source: ECLAC/CEPALSTAT.

Database for Latin America and the Caribbean (CEDLAS and the World Bank), the GINI coefficient in Argentina increased in the 1990s from 0.452 (1988) to 0.496 (1998), reaching 0.524 (2003); it started, since then, to decrease, reaching the level of 0.432 in 2010. In Brazil, the GINI coefficient has been decreasing gradually from 0.64 (1989) to 0.6 (1995), 0.588 (1998), 0.564 (2003), reaching 0.529 (2010). But the numbers that support an inflexion in the Brazilian case are the poverty rates. Argentina and Brazil are among the countries in Latin America that have led the trend in poverty reduction in the first decade of the twenty-first century. In the first period, poverty rates increased in Argentina 17 percentage points and in Brazil dropped one point; since 2003, poverty dropped by 25 and 15 points, respectively (Table 1.4).

Among the main factors to explain this reduction, scholars name the consistent growth in real minimum wages and the increase in social security programmes, among which are conditioned cash transfers (CCTs). In 2010, Argentina's CCT programmes Programa Nacional de Becas Estudiantiles and Asignación Universal por Hijo accounted for 0.2 per cent of GDP, and Brazilian CCT programme Bolsa Família accounted for 0.5 per cent of GDP. Argentinean programmes covered 8 per cent of the entire population and 46 per cent of the target, which were very poor; whereas the Brazilian programme covered 26 per cent of the population and 85 per cent of the targeted poor. The figures are very different; the coverage and the expenditure of Brazilian programmes of CCT, the largest in the world, are much wider than the Argentinean one (Lavinas 2013).

Relying on such results, the governments from the second period analysed have built a discourse around their progressiveness in contrast to the conservative other. However, it is not agreed that it was an active state policy to increase the percentage of spending, or it was a mere effect of the economic growth, together with the unprecedented high levels of commodity prices, which increased state revenues (Valdés forthcoming). Pérez Sáinz (2015) warns against a deterministic understanding of the role of extractive industries in state revenues in Latin America: it is necessary to assess whether states increased their revenues by a highest share of rents or due to other policies. The latter point is especially relevant because the most effective policy to reduce inequalities – not only poverty – would have to target the tax system; whereas the increase in state revenues from primary exports is often prior to the sphere of production. Indeed, critiques to this new political context can be grouped into three debates: the economic vulnerability of the model, as economies are integrated in the world market as commodities exporters and thus very dependent on external demand

and price formation (Gonçalves 2012; Sampaio Jr 2012); the limits of poverty reduc-tion schemes that are not accompanied by a welfare regime for citizenship rights (Lavinas 2013); and the socio-environmental impacts of this developmental model as well as its political impacts due to the criminalization of social movements contesting extractive industries (Gudynas 2012; Svampa 2012).

Research data

This book draws on different types of data, including official databanks, documental sources, secondary literature and primary data collected in interviews, newspaper art-icles and on the Internet. Interview partners were selected from an exploratory research with media material (Motta 2013b) and complemented by snowballing, until reach-ing a process of saturation, when narratives overlapped. In Argentina, I interviewed representatives and leaders from peasant movements from the provinces of Santiago del Estero, Córdoba and Buenos Aires; national environmental and agroecological organizations; a group of neighbours affected by pesticide spraying and their allies forming a network of NGOs, doctors, scientists in Córdoba; a journalist who has been covering these struggles for many years in the newspaper *Página/12* and on alternative media; and a scientist who was a fundamental ally for these groups. I had the oppor-tunity to observe a court ruling over pesticides in Córdoba, on 7 August 2013, and an assembly of the National Peasant Indigenous Movement (MNCI) at Universidad Campesina, from 20 to 22 February 2013; in both occasions, many activists were gath-ered. In Brazil, interview partners included: national peasant movements; national environmental, agroecological, consumer rights and human rights organizations; as well as scientists who were important allies and worked with social movements (see Appendix).

I complemented the accounts from activists with campaign documents and with political communications that are published in the mass media and on the Internet; when possible, the data was compared with official databanks, media material and secondary literature. Media material was collected systematically for the period of 1997–2004 for *El Clarín* and 1995–2004 for *Folha de São Paulo*, the two leading news-papers in each country. I used keyword searches in the online archives.[11] The period for which data was collected offers the most revealing data, because it comprises the first years of the introduction of GM crops. In addition, the research draws on media data for the period of 2009–2010 from previous investigations by the author (Motta 2013b, 2015; Motta and Alasino 2013).

Notes

1 This section is an adapted version of Motta (2014), 'Social Disputes over GMOs: An Overview', published in *Sociology Compass* 8 (12): 1360–76.
2 It was coined to study the mediation between raw material supplies to final con-sumption, including processing trading, wholesale and retail. Such connections shed light on how value is constructed, which might happen at different geograph-ical locations, integrating them into asymmetrical positions in the global division of labour of the world economy (Hopkins and Wallerstein 1986).

3 From a native perspective, it would appear inappropriate to use the analytical category 'food regime'. As will become clear in the chapters to follow, one of the
main criticisms from activists is that this regime is not about producing food but
commodities. When they use the word 'food', they mean exactly the alternative to
this process of commodification; they mean it as the production of something that
nourishes and which is sacred to the extent that it should not be tainted by the global market logics being applied to commodities.

4 The source of these numbers is highly problematic. James is the founder of the
International Service for the Acquisition of Agri-biotech Applications (ISAAA),
which has strong interests in inflating the numbers. The biotech industry is behind
it and recurrently uses the argument that the wide adoption of GM crops proves its
benefits to farmers. The second reason is that their data is indeed a point for objection: they calculate the area of crop by genetic trait, so a crop with stacked-genes
is counted as many times as the traits it contains (The Africa Centre for Biosafety
2013). Nevertheless, it is the most quoted data, not least because of their successful
media relations.

5 The universe of positive cases would be Brazil, India and South Africa (Schurman
and Munro 2010; Scoones 2008). Recently, mobilization has been increasing in the
USA, Uruguay and Paraguay.

6 Other negative cases would be the USA and Canada, where the resistance never
threatened the interests of the pro-GMO coalition (Pechlaner and Otero 2008;
Schurman and Munro 2010). In China, the state policy on GMOs is not aligned
with the interests of the big biotech companies, but the role of social disputes was
less decisive than the state autonomy (Newell 2006). The size of Uruguay and
Paraguay and the strong influence of Argentinean and Brazilian agribusiness in
these countries (Fassi 2009) would make them difficult to compare to the bigger
countries. There is not much research on Pakistan; it could be an interesting case
to compare to India.

7 In more general terms, they are geographically contiguous, have similar development indicators, and also share many commonalities regarding colonial legacies,
political systems and recent histories: dictatorships in similar periods in the twentieth century and a parallel democratic transition starting in the 1980s; they are
the founding partners of the regional integration bloc MERCOSUR; both experienced a neoliberal decade in the 1990s, ending with a change in power, when for
the first time since the return to democracy, governments with a left-wing discourse
were elected.

8 Following the economic crisis that hit the overvalued currency both in Argentina
and Brazil, a rupture occurred from the financial state that characterized the 1990s
(high interest rates, liberalization of markets to international competition, parity
in exchange rates, high levels of imports, and all financed by privatization and
public debt).

9 Since 1996, the *Kandir Law* exempted exports of agrarian and semi-processed
products from various taxes. Since then, proposals in this direction have been
introduced, but did not succeed, and the Ministry of Agriculture positioned itself
openly against it (Franco 2014).

10 There are three main taxes: rural property, production and commercialization. The
rural land tax (*Imposto sobre a Propriedade Territorial Rural*, ITR) is charged by

subnational states on the reported value of production. The second type is social contributions based on gross revenues (including financial) of the agrarian production, if the producer is formalized as a corporate entity. They are taxed by the federal union and the rates vary from 3–7.6 per cent (PIS/PASEP) and 0.65–1.65 per cent (Cofins). The taxes over commercialization include the Contribution to Rural Social Security (CERSS) of 2 per cent, charged by the federal union, and the Tax on the Circulation of Merchandise and Interstate and Intermunicipal Transportation Services and Communications (ICMS), charged by subnational states, which distributes a percentage to municipalities. It charges the movement of products for domestic sale or processing and varies according to states (Lazzarotto and Roessing 2004). Since 2013, the federal union enforced a tax exemption of 9.25 per cent PIS/Cofins (art. 29 Act 12,865). Some states also exempted soy products from paying ICMS.

11 This offered both advantages and limitations. The main limitations were that content from *El Clarín* started to be available in the online archive only from mid-1997 onwards; and there were many errors in the archive for the year 2004. This is also why the search was limited to the year 2004, because the following two years had many links missing in the search results. The great advantage was the possibility of making an efficient search.

Chapter 2
A silenced revolution (1996–2002)

During the early 1990s, institutional reforms paved the way for neoliberal restructuring of agriculture and for the approval of genetically modified soy. Those directly affected by the expansion of a new agrarian model – like small farmers, traditional communities living in the countryside and inhabitants of semi-urban neighbourhoods – first had to interpret what was going on. The transnational organization Genetic Resources Action International (GRAIN) was better equipped with information, due to its presence in transnational advocacy networks, which had already been part of the long process of elaborating on the meanings of the new technology. When GM soy was approved, there were two different contexts of reception of the new technology: the countryside and the urban centres. This first phase culminated with the financial and political crisis of 2001, which also marked a cycle of contention in the Argentinean society. These are the main events of the development of bio-hegemony in Argentina, which will be told in this chapter. Figure 2.1 summarizes these episodes.

The political economy of neoliberal adjustments (1990–1996)

The 1990s became known as the decade of neoliberal reforms in Argentina, headed by President Carlos Menem (1989–1999). He was elected by the Justicialist Party (PJ), the official party of the Argentinean political movement that became known as *peronismo*. It is the biggest party in Argentina, and the one with widespread mass support due to the legacy of the social and workers' rights that Perón's government recognized. In 1994, Menem passed a constitutional reform that allowed for re-election, which he profited from as he was elected for another term. This was only one of the several reforms that his government undertook.

Already in the late 1970s, during the military regime, Argentina adopted some reforms of market deregulation and, at the end of the 1980s, it advanced trade liberalization, culminating in 1991 with the elimination of quotes, the reduction of tariffs (from 30 per cent to 18 per cent) and the removal of most taxes on exports. Due to its speed and scope, Trigo *et al.* (2002) describe this process as drastic: there were no alleviating measures to assist national firms in preparing for foreign competition. The market opening affected capital goods in particular.[1] Argentina pursued an overarching privatization programme; issued licences for exploration of natural minerals and energy; and deregulated markets to allow for the entry of foreign capital in many sectors, including services. These reforms were embedded in an international climate of neoliberal economic reforms that took a specific form in the disciplines of the World Trade Organization (WTO) agreements, signed in 1994 in the Uruguay Round.

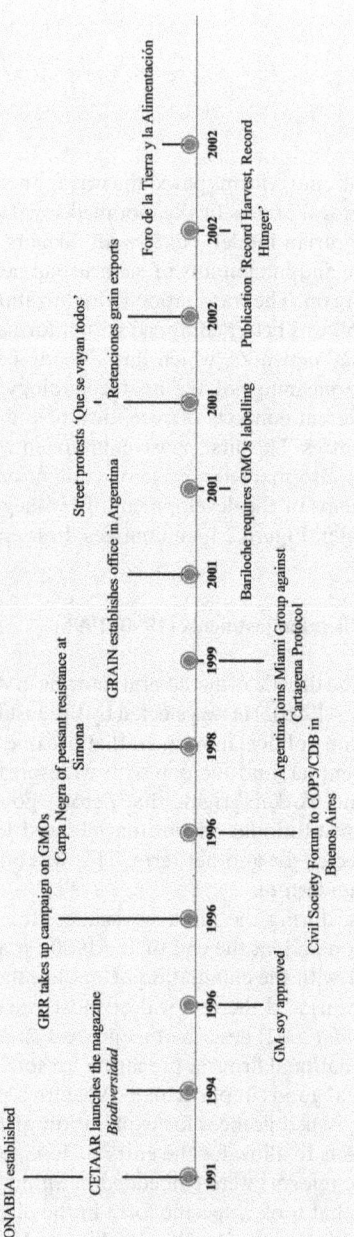

Figure 2.1   Timeline of events in social mobilization against GM crops in Argentina (1991–2002)
Source: Own creation.

The reforms dictated by the Washington Consensus did not spare the countryside and most instruments of agrarian policy – such as public funding, price control and public regulation of stocks – were abandoned as part of structural adjustments. This left the sector to the market forces of competition and only larger farmers and those with access to private credits were able to make a good transition to the new context. The agrarian sector presented a steady growth in the 1990s, even when the GDP contracted in 1995. There was a significant increase in the planted area, in the volume of production and in the volume of exports (favoured by the elimination of export duties). Exports increased in grains and oilseeds in the Pampa region. At the same time, the sector witnessed a process of fusions and acquisitions. Trigo *et al.* (2002) highlight two immediate causes: the expansion of agricultural land at the expense of cattle farming and the increase in the productivity of yields. These were affected by microeconomic factors (such as prices influencing the decisions of farmers), by the macroeconomic conditions, such as the cuts in tariffs for capital goods, as well as by a 'technological change', which includes the intensification in the use of capital goods, machinery, fertilizers, agrochemicals and a particularly important change: the introduction of transgenic seeds (Trigo *et al.* 2002).

However, neoliberalism did not mean the absence of the state. The state had a central role in fostering biotechnology by establishing the institutional conditions that gave the legal certainty required for investments in the country, as transnational seed corporations were looking for places in the southern continent where they could conduct off-season field trials (Burachik and Traynor 2002; Pellegrini 2013). In fact, Argentina had the highest number of field trials for GM crops in the beginning of the decade (James and Krattiger 1996).

In 1991, the Argentinean state had already created a body to regulate agrobiotechnology, the National Commission of Agrarian Biotechnology (CONABIA). Under the auspices of the Secretary of Agriculture, Livestock, Fisheries and Food (SAGPyA), located at the Ministry of Economy, the commission was to be composed of experts from both the state and industry. CONABIA's scientific profile not only aimed at the insulation from political considerations but it also created political consensus in the choice of scientists (for their clear career profile of pro-agrobiotechnology), dismissing the need for an exchange of scientific arguments. In CONABIA, a policy culture based on expertise that did not support scientific controversies was established and was not open to public scrutiny or actors who could raise critical objections to biotechnology (Poth 2013).

The *Seed Law* (Act 20,247) of 1973 established a national register of plant varieties to ensure commercial protection. Argentina only formally adopted the treaty from the *International Union for the Protection of New Varieties of Plants* (UPOV) in 1994 (Act 24,376), in the context of the negotiations concerning the WTO agreements. Having opted for UPOV 1978, Argentina guaranteed breeders' exemption and farmers' rights to save seeds. It recognized the breeders' right for commercial protection, but opted not to confer patent rights over plants under the legal framework of intellectual property rights, and rather offered commercial protection for breeders under the seed law. Genes are submitted to patent, but not plant varieties, which means that the breeder has commercial protection together with the proprietor of the genetic modification (Arza *et al.* 2010). This means that farmers' rights are guaranteed in a more extensive way, as in the case of patents (Rauchecker 2013).[2]

Nevertheless, there were actors who resisted the constitution of a new agrarian model, based on neoliberal global regulations and new technologies. Indeed, the expansion of the bio-hegemony project was one important driver to reawaken the peasantry as a political subject.

The retrospective view from the rural poor: technological and social change

At that time, there was not an organized peasant movement as there is today.[3] Nowadays, the members of the MNCI have a very structured narrative of the period that they call 'the transgenic revolution'. Recalling those times, a leader of the Córdoba Peasant Movement (MCC) separates the story of their resistance to the new agrarian model into two moments: in the 1990s and from 2005 on. The first moment was characterized by the low level of organization of the peasant movement and by a belated stage of the Green Revolution in Argentina.[4] To the activist, these two aspects are interrelated: if the advancement of the Green Revolution was so tenacious, it was in part due to the absence of agrarian movements in the countryside; at the same time, it motivated popular sectors in the countryside to organize themselves locally:

> [T]he introduction of transgenics takes place here in a context where the peasant movement was neither strong nor articulated yet. There were organizational processes locally of certain struggles, but there was no strong political coordination and in this context the new green revolution advances.
>
> (interview with MCC, 2013)

In his view, this Green Revolution involved not only new technologies, but also larger social, economic and political aspects. Economically, there was a change in the meaning of land towards a sheer commodity and a stronger orientation towards monoculture destined for export markets. Social transformations denoted incipient signs of a possible alliance between two different social classes in Argentinean agriculture: the landlords (*terratenientes*), part of the agrarian elite whose origins date back to colonial times, and the family farmers (*chacareros*) also known as *gringos* for being descendants from the European immigrants from the late nineteenth and the beginning of the twentieth century, who owned small or medium-sized lands. The landlords are represented by the powerful Argentinean Rural Society (SRA). The family farmers are represented by the Argentine Agrarian Federation (FAA). Actually, a political alliance between them would only happen in 2008, in the context of a conflict between agrarian actors and the government. But it was only possible due to the agrarian and class transformations taking place since the 1990s.

The transformations in family farming have been documented by Gras (2009).[5] One trend was the decreasing number of farm units, as many bankruptcies forced farmers to exit the market by selling or renting their lands and many lands were auctioned off as debt payment. Those who were able to continue farming did so by capitalizing, incorporating technology and leasing land to increase the scale of operation. These became family entrepreneurs as farming took on the characteristics of a business: the production process changed, with the use of larger machinery for sowing, spraying pesticides and harvesting, often by subcontracting large enterprises of machinery services. This decreased the importance of family-based labour, releasing new generations from the tradition of becoming farmers themselves and they were sent to study in the urban universities to become farm managers.

Changes in land tenure showed a tendency for productive concentration rather than concentration of land ownership. This means that non-farmers entered into the agrarian sector seeking profits through sowing pools. These new actors invest capital to lease land in different places to reduce risk, in which agronomists manage production by hiring machinery firms. With more capital to rent lands, there was an increase in land prices, pushing family farmers harder into either being able to compete and increase scale, or opt out. Gras (2009) highlights the role of state intervention since the early twentieth century in creating the conditions of coexistence between large-scale and family farming. The reforms introduced in the 1990s lifted the regulations that had been fundamental for the family farming model in Argentina and eroded its viability.

In a new paradigm in which land ownership lost its political, economic and symbolic power, and land is considered to be a mere productive factor, the power of landowners from SRA diminished, and the material as well as the symbolic reproduction of family farming too. The new generation of family entrepreneurs was trained to be managers, socialized in universities more than in the fields, in a process in which agriculture is understood as business and disconnected from rurality (Gras 2009). Gras and Hernández (2008) claim that they are becoming a new ascending bourgeoisie, an aspiring leadership in the agriculture sector and hegemonic in society. Represented by business associations and technical chambers, they have made themselves distinct from the traditional landed-elites, represented in the SRA, while the FAA have undergone a reconfiguration in their membership.

Peasant movements interpret the processes taking place in the 1990s as an offensive against peasant farming. It included both a discursive strategy to build a hegemonic conception of agriculture as agribusiness as well as material and institutional means, as this conception was promoted by public policies:

> So when the transgenic revolution takes place, there is already a very big hegemony of this conception of agriculture and a popular sector very weakened because we came from the process of military dictatorship and we were in the heyday of neoliberalism.
>
> (interview with MCC, 2013)

In short, in the retrospective view of the organized rural poor, the 1990s were a decade characterized by the formation of agribusiness and by a lack of organization and mobilization of popular sectors. Nevertheless, the picture is more nuanced than that. In fact, the 1990s were a moment in which solidarity and cooperation in civil society were undermined by neoliberal policies. Deregulation of the labour market weakened unions, and popular sectors were hit by precarization (Petras and Veltmeyer 2010; Svampa 2008). However, although there was not widespread social mobilization until the outbreak of 2001, the latter was 'but the crest of sustained anti-neoliberal protest' (Silva 2009, 56). There were protests at the local level, exposing new repertoires of contention, such as sieges of public buildings, barricades on roads and camps in central squares (Auyero 2004). In the mid-1990s, some groups of *piqueteros* (picketers) in the suburbs of Buenos Aires and in popular neighbourhoods started local networks, forming new social ties in order to survive the hardships (Svampa 2008).

This process also took place in rural sites: in 1995, the Movement of Agrarian Women in Fight (MML) emerged from the reaction of a woman who defended her family's small landholding from a warrant of execution of debts (Bidaseca 2003; Giarraca 2001).[6] She went to the local radio and called for support, and other women living in the same situation joined in protest to stop the execution of the debt. As authorities failed to deliver promised political solutions, the MML spread across the country, forming a network of assemblies. They discussed causes and solutions for their problems, reaching the conclusion that the generalized indebtedness was attributable to the government, which while encouraging them to capitalize their production lacked adequate policies to support them. The framing of the situation showed gendered differences: while the fear of losing their lands and the feeling of injustice motivated them to undertake collective action, often their men would take on the individual responsibility for the debt. Their gender identity also led to the choice of the International Women's Day to occupy the Plaza de Mayo in 1996 and 1997. By then, they had managed to suspend many warrants of execution of lands by protesting.

The MML innovated the repertoire of agrarian actors, until then institutionalized in unions such as the FAA, the local entities of the Argentinean Rural Confederation (CRA) and in the Federation of Cooperatives (CONINAGRO). Their fathers and husbands were often members of these organizations but the women from MML considered them patriarchal and disagreed with the FAA, understanding that small farmers should follow the path of technological modernization, while for them this was the reason for the indebtedness; thus they criticized the FAA's politics of negotiating with the government and forming alliances with large farmers. The MML, by contrast, opted for disruptive action and alliances with other social movements, particularly the (urban-based) feminist movement, with organizations of small business and farms indebted, with the Landless Workers' Movement (MST) in Brazil and with the Peasant Movement of Santiago del Estero (MOCASE). A dissident faction of the FAA, Chacareros Federados (Federated Farmers), was more respectful of gender issues, and thus became an ally too. The MML called into question not only the legitimacy of the indebtedness, but also the economic model and the absence of policies to support small farmers. For Giarraca (2001, 130), the MML represents 'an attempt to generate a new social subject, to produce new meanings to deal with a conflict that, beyond the punctual demand ... denotes a cultural problem ...: the possibility or impossibility of keep living in the countryside'.

Precursors (1990–1996)

At that time, there were already small NGOs working with issues of interest to small farmers and agrobiodiversity. Among these, the Centre for Studies on Appropriate Technologies in Argentina (CETAAR) had been interested in genetically modified organisms in its early phase.[7] Since September 1994, it started distributing the magazine *Biodiversidad*, launched in that same year by the transnational NGO GRAIN, whose founder, Henk Hobbelink, was one of the pioneers in the global movement against GM crops (Schurman and Munro 2010). Thus, before the approval of GM seeds in Argentina, CETAAR had started calling attention to the issue, publishing articles in *Biodiversidad* with a critical perspective on agrobiotechnology (interview with GRAIN Argentina, 2013). CETAAR provided the initial support for the later establishment of GRAIN in Argentina in 2001.

In 1996, Buenos Aires hosted the Third Meeting of the Conference of the Parties to the Convention on Biological Diversity (COP 3/CDB). The conference, Issues Related to Biosafety, approached the negotiations of what would later be the *Cartagena Protocol*. The convening of this meeting shows the active involvement of the Argentinean government in shaping the global framework for regulating GM crops. CETAAR, GRAIN and Via Campesina organized a parallel civil society forum, Los Pueblos en la Convención (Peoples in the Convention), to discuss GM crops (interview with GRAIN Argentina, 2013).

The approval of GM soy and initial reactions (1996–1998)

In 1996, CONABIA approved the commercial use of GM soy. A leader from the MCC describes the approval as quick, without public discussion and without health risk assessments and scientific considerations. 'This is how the legal processes went unnoticed', he adds (interview with MCC, 2013). While clearly invoking the role of science to serve as the basis for decision-making over biotech crops, he also calls for public accountability of this decision procedure. In other words, there is no anti-scientific position on the part of the peasant movement, but rather a demand that the policy culture rely on science under democratic and transparent conditions.

The activist presents a critical diagnosis of the situation of scientific institutions, also affected by the neoliberal agenda. Left without public funding to further their research activities, these were taken over by biotechnological firms, like Monsanto, owner of the Roundup Ready (RR) soy that had been approved. Science is placed in quotation marks as a discursive means to sharpen a critical view: '[U]niversities were looking for research funds from private sources ... thus many agronomy faculties begin to do research according to their contributors requested ... and it begins a whole trend of "science", in brackets, to create a biased scientific production, for the promotion of GMOs' (interview with MCC, 2013).

In his narrative, he identifies the processes by which corporations, in building bio-hegemony, acquired material power by financing universities and transforming scientists into their spokespersons. In addition, the activist identifies the institutional power of the pro-GMO coalition by highlighting two types of institutions that played an active role in disseminating the technology: the National Institute of Agricultural Technology (INTA)[8] and private actors such as technical chambers:

> Immediately, a strong dissemination and publicity on the virtues of GMOs begin, through state institutions first. The INTA plays an important role at this moment; the technological modernization is already installed in this school, of associating technology to reduction in labour costs and increase in production.
>
> (interview with MCC, 2013)

In the private sector, the Argentinean No-Till Farmers Association (AAPRESID) institutionalized a coalition formed of members from the various sectorial representative associations in Argentina and transnational corporations. The machinery for no-till farming is part of the technological package composed of GM seeds and glyphosate and was widely propagated as being environmentally friendly. The role of AAPRESID cannot be overstated (Hernández 2007). They were facing a problem with pests that prevented them from expanding their business with no-till farming machinery. Monsanto's total herbicide offered them a powerful solution. Therefore, the association of their business, no-till farming, with the technological package GM seeds-glyphosate, positioned them as leaders in the launching of the new technology.

Although there has been no public debate around the decision to approve GM soy, a former activist from Greenpeace (interview with former campaigner Greenpeace Argentina, 2012) recalls that it was not difficult to become aware of the issue, because the technology was advertised by seed companies as a big solution for the country. Greenpeace, moreover, is an international organization and part of an environmental transnational advocacy network. In 1995, before GM seeds arrived in Argentina, the international assembly of that organization had already decided to take up a global campaign against GMOs. The decision of participating in a global campaign had to balance between local priorities and international interests. It should also follow the rules established by the international director, namely campaigns should only be pursued if goals have a possibility of being achieved. Such rules have influenced the Argentinean decision not to create their own national campaign, although they took up some actions, as they evaluated at the time that they had no chances of winning. In 1996, GMOs started as a topic under the heading of the 'biodiversity campaign'. Today, GMOs are listed as a subtopic of the campaign on forests (*bosques*). The narrative under the heading of *Transgénicos* on their website reveals the involvement of Greenpeace with the issue:

> In Argentina we consider that, after more than a decade of campaigning, the enactment of the Forest Act in late 2007 was, in addition to a protection to vital ecosystems such as native forests, the most effective deterrent achieved to stop the expansion of GM soy in the country.
>
> (Greenpeace Argentina 2013)

Greenpeace gives a number of reasons for the difficulty they faced initiating a public debate on the issue when GM soy was approved. The advertisement of the new technology did not warn of any negative impacts of the new technology. At that time, 'the direct impacts were not tangible' (interview with Greenpeace Argentina, 2012). This is one reason that made a campaign against GMOs a difficult task: there was a lack of empirical credibility of claims against it. A second reason was the lack of organizational capacity at that time, and third, the low resonance of environmental concerns in the Argentinean society. Concomitantly, due to the structural logics of mass media, there was less media space for covering environmental topics: 'The environmental organizations were smaller; environmentalism … did not exist in Argentina in the nineties … so it did not generate interest on the topic in society, because it was not in the mass media' (interview with Greenpeace Argentina, 2012). The low resonance of environmental issues with the public and in the mass media must be contextualized in the zeitgeist: 'The society embarked in this project of neoliberal thinking, with all these ideas that everything that is imported is good and all the rest is left non-discussed' (interview with Greenpeace Argentina, 2012).

The initial decision of Greenpeace of not taking up a campaign against the approval of GM soy in Argentina was the point for Group of Rural Reflexion (GRR) to act. Their members came from the Peronist youth, many had been exiled, and since the 1990s they had been working in the Ministry of Agriculture. Even though they worked where RR soy was approved, they did not become aware of it through internal channels. They first heard of the new technology by reading foreign bibliographies and they heard of its arrival in Argentina when a journalist told them that Greenpeace had made a survey on the acceptance of GMOs, concluding that there was no room for a campaign against it:

> If this is not a topic for Greenpeace, so it must be for us. So we contacted European friends and started to anticipate the model of country that would come out of the industrial and technological package ... based on the assumption that agriculture is agribusiness.
>
> (interview with GRR, 2012)

By participating in transnational networks where the debate on industrial farming was taking place, the GRR started to write documents on issues such as agrifood chains, supermarkets and what was under construction at that very moment: agribusiness and the integration of Argentina in the world agrifood system as a provider of cheap commodities for food industry and animal industrial farming.

Over the span of four years, 90 per cent of the soy was GM. Argentina soon became the second world producer of GM crops. Some of the activists interviewed suspected that the government distributed seeds as an explanation for the rapid adoption (interview with GRR, 2012; interview with Greenpeace Argentina, 2012). There were also rumours that farmers' associations were involved in an agreement with Monsanto and INTA to multiply the seeds without having to pay royalties for them (interview with GRR, 2012). This has been the singularity of the Argentinean case in the promotion of GM seeds and remains a puzzle for many analysts that try to understand why Monsanto did not file the patent for RR soy. The GRR offers the explanation that Monsanto was interested in profiting from the product Roundup, not with the seeds (interview with GRR, 2012). Others have a different hypothesis: Monsanto would have reached the agreement with SRA not to file a patent in order not to generate any debate regarding the new seeds (interview with MCC, 2013). In view of the heated disagreements on a royalty system that ensued more than ten years later in Argentina, this explanation seems plausible (Seifert 2012).

The lack of a patent and of a system of enforcement of intellectual property rights partially explains the diffusion of GM seeds, together with the lower cost of the technology and the clandestine multiplication of GM seeds, which became known as *bolsa blanca* (white bag) because the packaging was not certified. This allowed for a structural change in the agrarian sector (Bisang *et al.* 2008; Varesi 2010), as the technological package increased the scale of operations; more and more lands were incorporated to plant soy, displacing other crops and cattle ranching (Catacora-Vargas *et al.* 2012). The modernization processes taking place in the Pampas in the early 1990s were replicated into neighbouring lands, setting in motion a fast expansion of the agrarian frontier (Pengue 2005). There was social mobilization against the advancement of the agrobiotechnological model in two different settings: in the city and in the countryside.

In the countryside (1999–2001)

The advancement of GM soy crops reached territories characterized by another way of occupation. Distant from the fertile nucleus of the *pampa húmeda* (humid pampa), these lands guarded native vegetation and were farmed by those who identified them-selves as peasants and indigenous peasants and did not have any legal titles as owners of the land where they lived for over a generation:

> [T]he peasant movement, in part, emerges in a situation of resistance, to stop evic-tions, to curb deforestation … In other words, the soy conflict in one place is trans-lated … to other regions, and the axis was always, it is, a situation in which there are peasant communities with no titles.
>
> (interview with MCC, 2013)

However, the existence of common grievances does not automatically lead to social mobilization. There must be a process of framing these structural changes in a manner conducive to collective action (Snow *et al.* 1986). It was only with their organization that communities of the rural poor managed to interpret these situations beyond an individual grievance and in terms of a social injustice. They became aware that they had the legal right to their lands – according to the law (*ley veinteñal*) that entitles those living on a land for more than 20 years to the right to its use. However, as any law, it was not applied automatically and the rural poor lacked the economic and professional resources to access the judicial system (Barbetta 2009). This was the moment in which the construction of the collective identity 'peasant' played a cen-tral role (Figurelli 2013). Their identity was framed in opposition to those identified with agribusiness, with an acknowledgement that they were in an unequal situation to dispute with them as their only resource was to act collectively and fight (*luchar*). Moreover, they perceived that by acting together they could make a difference. Years later, they are proud of not having had any case of an evicted peasant in the move-ment (interview with MOCASE-VC, 2013). The success of their collective action can be understood in the combination of the elements of identity, agency and injustice framing (Gamson 1992).

GM soy became thus an issue for the peasant movement already combined with the land issue: 'When the issue of GMOs started? I believe that we started to bring the issue, easily, from 1999 or so, already hard, when we were living the issue of evictions' (interview with MCC, 2013). They resorted to a new action repertoire, borrowed from the incipient protests taking momentum in the cities as a reaction to the neoliberal reforms: the white tents from a teacher's protest (Auyero 2004).

> A fact, say, historic, of the resistance was the *carpa negra* at Simona … A tent of the entire community was settled there … resist … the advancement of bulldozers. And it was … a modality that after multiplied, the black tent of resistance, the black tent of land.
>
> (interview with MCC, 2013)

In 1998, the black tents (*la carpa negra*) in the locality of Simona, in the province of Santiago del Estero, meant more than the diffusion of a repertoire of action from the city to the countryside, which later became modular. It also signalled the establishment of a network connecting both sites of contention against a common enemy: the neo-liberal reforms.

The protests organized by teachers, workers and the unemployed in those years were the forerunners of a cycle of contention against neoliberalism. The unemployed formed a movement that became known as *piqueteros* (picketers), which would become prominent actors in the new cycle. The peasant movement established a sustained cooperation with the *piqueteros*, the Argentine Workers' Central Union (CTA) and the Coordinator of Popular Autonomous Organizations (COPA): 'it was no longer a peasant question, who is evicted, it became also an issue of the agenda of resistance to neoliberalism' (interview with MCC, 2013). Musicians came to Simona to demon-strate their solidarity and, when playing at important festivals, invited on the stage someone who had been the victim of forced evictions. The direct testimonial not only resonated with the public but also offered good coverage for the media (interview with MCC, 2013). Also the MST sent representatives from Brazil to express their solidarity.

This moment marked the beginning of networking with two urban actors: per-ipheries (*barrios*) and university students and researchers, in particular the Argentina Federation of Students of Agronomy (FAEA). They organized joint activities such as roundtables and conferences. Representatives from the peasant movement came to events at the university and vice versa. The cooperation with urban actors took the shape of an exchange and internship programme that brings students and youth from the *barrios* to live in the countryside within the peasant communities; then, youth from the countryside also stay in the *barrios* for the period of a month. All these activities helped to spread the peasant movement agenda (interview with MCC, 2013).

On the fringes of the city

The contact with the countryside was also a goal of the urban organizations that had been active in issues of food production and distribution. This was the case of Service to the Popular Culture (SERCUPO), which later became the only urban organization member of MNCI. Their history dates back to the period of high inflation at the end of the 1980s when they started taking part in the alternative markets at the *barriadas*[9] as palliative measures to cope with inflation. This is part of a process that movement scholars call the *territorialización* of collective action in Argentina to highlight ini-tiatives that took place at the local level, in the places where people lived, as collect-ive action among workers had receded with the precarization of the labour market (Luzzi 2012; Svampa 2008). The idea of a popular culture for SERCUPO is to draw on communitarian means of organizing the economy to fight poverty. They recall having started their contacts with MOCASE around the year 2000 when they first learned of the peasant question and established the exchange and internship programmes.

The frame bridging with the peasant movement was facilitated because most inhabitants of the *barrios* or their parents had once been peasants and had to leave their lands due to the structural transformations. Memory and identity construction were also part of their meaning-building work: 'According to the time of existence of the barrio, one feels it a lot ... Boys hunting animals, families with orchards, there is a lot of this that is still preserved and one must, say, value it, rebuilt it, not let it lose' (interview with SERCUPO, 2013). The crisis of the 1999–2001 period provided some of the conditions for advancing their work of promoting a community-based food production. SERCUPO made alliances with the movement of recovered factories, with a meat processing plant in particular, in whose lands they started cropping (interview with SERCUPO, 2013).

Framing soy to contest neoliberalism

GM soy became a very concrete symbol of protests against the impacts of neoliberal reforms in the countryside, which can be seen with the protest signs at the *carpa negra* in Simona (Mohaded 2000). Soybeans were the visible face of the transformation in Argentinean agriculture. Peasant movements started an interpretative work on the causes of their grievances and their targets of responsibility. The diagnostic framing included the increasing greed for land and the use of a new technological package composed of GM seeds, new machinery and pesticides. The attribution of responsibility is complex: it lies in the economic power, but also in its allies in politics and in the media (interview with MOCASE-VC, 2013).

On the one hand, the diagnostic framing of what was going on was straightforward: given their experience of forced evictions, deforestation and pesticides spraying, the frames of land rights as well as health and environmental protection had resonance for them due to the experiential commensurability. It was not easy though to understand what *Roundup* or RR (the product developed by Monsanto based on glyphosate to be used in association with the RR soy) was, what genetic engineering was, and how RR could kill all the weeds and not the soy plants:

> [F]irst one hear RR, you do not know ... We always said that with the entire weed that it destroys, that it would be impossible ... When they spray, that is, what a power this plant must have to resist, because all the rest dies. Then we started researching ...
>
> (interview with MOCASE-VC, 2013)

On the other hand, the framing of the agrarian model was facilitated because since 1994 regional organizations like MOCASE had started participating in the activities of Via Campesina: '[T]his also goes hand in hand with our participation at Via Campesina. It brought elements to build a clearer view of the model and start denouncing the issue of GMOs and pesticides' (interview with MCC, 2013).

The agronomy students taking part at the MNCI also helped to explain how the expansion of GM soy brought violence to their territories, which had until then no commercial value, by integrating them into the land market. In hindsight, it is easier to provide an explanation to what happened, but the interviewees share the view that this is a result of their work of interpreting and giving meaning to it as they suffered the consequences of the process; no interpretation was readily available at the time to help explain the lack of organized resistance: 'But, basically, during all the first stage there was no organized resistance or discussion in terms of whether GMOs were good or bad. Today, perhaps, this is given, but now with 20 million hectares of GM soy' (interview with MCC, 2013).

In the cities: targeting urban consumers

Faced with the fact that Argentina became the second largest world producer of GM crops, Greenpeace organized some actions to raise public attention to the issue.

Aiming at the approval of labelling rules, they framed their message in terms of consumer rights and health. They relied on the division between what happens in the agrarian fields and what happens in the supermarket with urban consumers:

> [O]ne cannot say that, because Argentina and Brazil plant a lot of soybeans, Argentineans and Brazilians eat GMOs and do not care. That's not a correct statement, because one thing is what happens on the field, the other thing is what happens in the supermarket. These are two different worlds.
>
> (interview with former campaigner Greenpeace
> Argentina, 2012)

Greenpeace disseminated information on the issue and organized media-designed protest events. One of the protests involved Frankenstein's monster cooking a large pot of soup made with Knorr bouillon cubes that contained GM ingredients. People came to have their plates filled, blindfolded, without knowing what they would eat (interview with Greenpeace Argentina, 2012). However, campaigners name major obstacles in mobilizing people for the issue. First of all, more than 90 per cent of the GM crops cultivated in Argentina were exported. Even the soy that was destined for the internal market was mostly for food processing or animal feeding. This means that consumers did not directly eat GM products. In other words, the genetic engineering of RR soy was situated in the initial stages of a global commodity chain that included many processing stages, making it difficult for consumers to visualize a GMO, especially in a context of transformations in food habits as people started to rely more on processed and industrial food, a constitutive element of the corporate food regime. Thus, Greenpeace found it difficult to provide empirical credibility to the claims regarding consumers' health:

> It has been proven that where labelling was introduced, people prefer not to consume it [GMOs] ... But if it is not consumed, as it happened in 1996, and it is exported, if some chicken in China eat it, why should people worry?
>
> (interview with Greenpeace Argentina, 2012)

The former Greenpeace campaigner recalls that there was a lack of scientific studies that proved risks, an important resource for mobilizing against GM crops. This shows also the dangers of accepting the rules of the game of regulatory policy based on science, given the dominance of mainstream science, not permeated by environmental and societal concerns. The activist calls for an epistemological turn in science, in which the precautionary principle guides the application of science into politics. To him, science should call into question the concept of substantial equivalence, which serves as a basis for the regulation on GMOs in the USA and Argentina and search instead for the differences that might result from the mechanisms of genetic insertion (interview with former campaigner Greenpeace Argentina, 2012).

The mobilization efforts nevertheless showed some results. Protest actions from Greenpeace in Europe against GM imports from Argentina provoked the reaction of AAPRESID and were covered in the media (Trucco 2001). Then, Greenpeace Argentina published an article in *El Clarín*, explaining its position (Ezcurra 2001). Political opportunities were found in the federalist structure of the Argentinean state and activists achieved a local victory in 2001, with a municipal ordinance from Bariloche establishing obligatory labelling (Cordero 2001). As authorities entered the judicial system to declare it unconstitutional, activists engaged in legal mobilization and involved the public defender's office. This grounded its defence of the ordinance in the right to information, justified in constitutional terms and in human rights treaties, as well as in arguments based on uncertainty and the precautionary principle. Finally, the document states that the approval of GM seeds in Argentina was a political decision and an illegal and unconstitutional act. As well founded as this argumentation was, it had no tangible effects. The governmental policy was unambiguously promoting GM technology. Without much resonance with the public, they had no chance of winning this battle. For an organization like Greenpeace, this implied that the issue would have to be dropped.[10]

Negotiating legal frameworks internationally and in the parliament (1998–2001)

During the negotiations of the *Cartagena Protocol*, Argentina left the Group of 77 (G-77) pressured by the USA and by Monsanto diplomacy and allied with Australia, Canada, Chile, the USA and Uruguay in the creation of the Miami Group (Newell 2009). The alliance fought against an international treaty and any commitment that would jeopardize the production and export of GMOs. The Miami Group tried to exclude from the scope of the agreement all GMOs destined for food and feed purposes and managed to keep them from the informed agreement procedure. Despite their attempts to exclude the precautionary principle, the *Cartagena Protocol* has a strong version of this principle of the environmental regime. Most exporters of GM crops did not sign the *Cartagena Protocol*. Argentina signed it, but it has not ratified it and thus has not incorporated its obligations, including the *Biosafety Bill*.

Between the end of 1999 and 2001, GRR, Greenpeace and other organizations tried to influence the legislative process with bill projects that foresaw the incorporation of the precautionary principle, labelling provisions and a larger oversight of the Ministry of Environment. Sponsors of these projects justified the need for Argentina to incorporate the obligations under the *Cartagena Protocol*. In 2001, there was another bill project, from the pro-GMO coalition, with the goal of turning the existing regulations into law. The activism of the GRR, Greenpeace and their allies helped prevent the project from being passed. Also, politicians aligned with the Ministry of Agriculture refused to debate the projects, which never came to be, for fear of a public debate that could have had unintended consequences for the future of the technology. In fact, the issue was left out of the legislative arena until 2008, when a law for the development of biotechnology was passed (Poth 2013).

These years coincided with a period of political and economic turbulence that culminated in the crisis of 2001. But before that, another event deserves to be mentioned: 2001 also marks the launch of GRAIN in Argentina. This was a result from a change in the profile of the organization in the direction of working closer with local movements in Africa, Asia and Latin America:

> It was when the peasant movements emerged, in the 1990s that GRAIN decided: if the peasant movements are the protagonists, what we have to do is to work with them. And then it started a process of decentralization, of working regionally with the movements.
>
> (interview with GRAIN Argentina, 2013)

Genetically modified crops play a prominent role in all the activities of the organization. However, GRAIN preferred not to engage in the campaign on labelling, not only due to its focus on social movements but also due to the perception that it was not effective where enforcement remains very weak (interview with GRAIN Argentina, 2013).

Crisis and transition: record harvest, record hunger (2001–2002)

The year 1999 marked the end of the government of President Carlos Menem. His party, the PJ, failed to get its candidate elected. The Alliance for Work, Justice and Education established between the parties Radical Civil Union (UCR) and Front for a Country in Solidarity (FrePaSo) was formed in 1997 and won the presidential elections in 1999, headed by the candidate Fernando de la Rúa. GRR organized a small protest action with Greenpeace and a group of consumer protection. They wore T-shirts with the writing 'No GMOs' and handed the president a plate with the slogan 'I don't want GMOs'. The president said he would consider the issue but that never happened (interview with GRR, 2012).

From 1999 to 2002, the Argentinean economy went through a serious crisis, after which inequality increased (Valdés forthcoming). The crisis culminated in mass protests, triggered by the announcement of restrictions to the withdrawal of cash from bank deposits, known as *Corralito*. There were food riots (Auyero and Moran 2007) in Rosario and Buenos Aires, and non-organized middle-class people went to the streets banging pots (*cacerolazo*) on 19 and 20 December 2001. As a reaction, the president declared a state of emergency; there was violence and deaths. The slogan *Que se vayan todos* (Away with them all) was heard as political leadership also became unstable (from 20 December 2001 to 2 January 2002, four presidents assumed office). The year of 2002 was a 'ambiguously extraordinary year' (Svampa 2008, 117), as it signalled both a deep crisis and a strong, mobilized society.

Meanwhile, the parliament nominated as president Senator Eduardo Duhalde, the candidate defeated by the PJ in the 1999 presidential elections. Among his many measures was the pesification (*pesificación*) of all dollarized deposits and the devaluation of currency. Agrarian policy became a new tool for financial stability (Varesi 2010). In April 2002, the president followed the IMF recommendation of raising export taxes (*retenciones*) on agrarian goods in order to finance public debt. The taxes would also be used to fund the president's unemployment scheme. Although this fiscal instrument had been commonplace, the Menem government had suspended its application during the 1990s. In fact, the increase in production of grains and their agrarian exports guaranteed an inflow of foreign currency and relieved state finances, as the agrarian sector was not in crisis like the rest of Argentina (Varesi 2010).

In June 2002, Greenpeace launched the report *Record Harvest, Record Hunger: Starving in GE Argentina*. The year 2001 marked a deep inflexion in poverty and access to food and, at the same time, an unprecedented volume of harvests. The gap between harvested soy and access to food lies in the commodification of soy and its insertion in global chains.[11] The report deconstructs the discourse of food security and instead defends food sovereignty. The chosen model of development, export-oriented and unjust, explains the distance between the two discourses (Greenpeace Argentina 2002). During that same year, 15 organizations[12] convened at the Earth and Food Forum (Foro de la Tierra y la Alimentación) and issued a document named *From the*

"World's Breadbasket" to the Soy Republic: Why We Are against the Transgenic Model.
In this document, they explained that the paradox of record harvest and record hunger had its root cause in the neoliberal globalization in the agri-food system, which they characterized as follows:

> It is a model dominated by the big transnational enterprises and the technologies that they control: the supermarkets in the final distribution of foods, the big food industry, the seeds and pesticide industry, and the concentrated financial capital (sowing pools).
>
> (Foro de la Tierra y la Alimentación 2002)

The NGOs deconstructed the paradox by making a clear distinction between commodities and food: as commodities are not food, there is no contradiction in the fact that Argentina simultaneously experienced a boom in commodities production and exports and a deterioration of food security: 'It is a model that produces raw materials (commodities) to export, and not food in enough quantity and quality for our population, forcing the import of various foods that used to be produced by our rural producers' (Foro de la Tierra y la Alimentación 2002).

Whereas this period shows the mobilization of the civil society against the agrarian model, the context was not receptive to their claims. Not only did the country rely on soy exports to equalize its balance of payments but the year also marked the beginning of a boom in commodities prices that would last until 2008. GM soy exports thus became a profitable source of state revenues.

Notes

1 In parallel to this international process of tariff reduction, Argentina and Brazil advanced their trade relations in the context of the regional integration project MERCOSUR.
2 Research has shown that small farmers do not use these rights because they cannot afford to buy seeds at all. The rules are beneficial for medium and big farmers; seed traders have been trying to restrict farmers' rights by making them sign a contract at the moment of the purchase, which forbids them to reuse the seeds, but farmers often do not comply with the terms of contract. The actual beneficiaries of the lack of enforcement of the seed law are the bigger farmers. As a consequence, seed producers have been pushing for a stronger regime and a new seed law in order to guarantee their share (Arza *et al.* 2010).
3 Incipient organizations started during the 1980s; before that, there were the *ligas agrarias* (agrarian leagues) during the 1970s that were part of armed resistance to the military dictatorship. However, their social composition was much more heterogeneous than peasants. I thank Carla Poth for those remarks.
4 Although the Green Revolution started in the 1970s with the military regime, many agrarian economists and other scholars also identify in the 1990s an intensification of those processes (Bisang *et al.* 2008; Trigo *et al.* 2002; Varesi 2010).
5 This and the next two paragraphs draw on her mentioned article, based on a fieldwork in the Pampa region.

6 This and the next paragraphs are based on these sources.
7 CETAAR represents Argentina at the Latin American office of the Pesticide Action Network (PAN), both established at the beginning of 1980s. This is a network of NGOs, farmers and research centres opposed to the massive use of pesticides and dedicated to denouncing their impacts and forming alternatives. Concomitantly, they object to GMOs.
8 However, there were disputes inside INTA and this institution was severely hit by budget cuts due to structural adjustments (Pellegrini 2013).
9 These are how the suburbs with precarious conditions for the urban poor are called, as well as the social and cultural activities taking place there.
10 Their action repertoire is designed to 'win' a policy change. Indeed, their website, under the heading of 'victories', collects about one or two victories per year with no mention of GMOs. There is even a gap in the years of 2000 and 2001, when the organization concentrated many efforts on the issue (www.greenpeace.org/argentina/es/sobre-nosotros/Nuestras-Victorias/).
11 When an initiative from agribusiness associations, the campaign Soja Solidaria, started marketing the consumption of soy as a nutritional solution to hunger, civil society organizations reacted by calling it an attempt to transform mono-culture into mono-consumption (Foro de la Tierra y la Alimentación 2002).
12 Argentina Association of Ethical Investigations, BIOS Argentina, Self-Summoned Ecofeminists, Environmental Protection Foundation (Funam), Ecological Informatics Network Foundation (RIE Foundation), Group of Rural Studies from University of Buenos Aires (UBA), Group of Agro-food Studies (UBA), GRR, GRAIN, Greenpeace, Institute for FairTrade and Responsible Consumption (ICECOR), Rainbow Initiative of Ecology and Society, Revista Futuros, National Ecological Action Network (Renace), Vida.

Chapter 3
Increasing noise (2003–2013)

After two years of crisis and instability, Argentina experienced a decade of relative political stability and economic growth. Genetically modified soy remained a central element of this path; however, this role was not unchallenged. This chapter analyses the processes of social contestation over the agrarian model between 2003 and 2013. It is divided into two parts. The first part covers the period from 2003 to 2008, starting with the new presidential government. Social movements disputing the agrarian model had to interpret the structure of political opportunities and had to look for, without success, spaces of influence. In 2008, a polarized conflict between agrarian elites and the government took over the national agenda for some months and established the political alignments between the agrarian sector and the government regarding the policy on genetically modified organisms. The second part of the chapter starts by analysing the consequences for the struggles of social movements fighting GM crops and the agrarian model. Social movements did not believe in the possibility of achieving changes at the national level. Mobilization, nevertheless, increased, targeting other levels while activists started to create new political spaces at the local level as well as within global and transnational arenas. Therefore, this chapter tells the story of how bio-hegemony was sustained as well as how it became increasingly the object of social mobilization. From a silenced 'transgenic revolution' as narrated in the previous chapter, the next phase shows much more 'noise' over the agrarian model even though it was not welcomed in the Argentinean national political system.

Fighting the other side of the agrarian boom (2003–2008)

Since the new government did not alter, but rather consolidated the GM policy, the movements resisting the agrarian model addressed the direct and indirect consequences of the expansion of GM fields. They often shifted the scale of their activities more and more to the local, regional and global levels. The main episodes in the disputes of the agrarian model during these first years of *kirchnerismo* will be described in seven steps, starting with an analysis of the ruptures and continuities of the new government, with an emphasis on its policy for GMOs. Greenpeace dropped its national anti-GMO activities, shifting the topic to the global level while concentrating its national efforts on forests. There were moments of networking and collaboration among the different organizations fighting GMOs, but these were confronted with difficulties in gaining allies among scientists and politicians. Meanwhile, the pro-GMO coalition ensured that the Argentinean state would engage in the promotion of agrobiotechnology in the multilateral arena. The peasant movement launched a national organization in 2005. Their struggles for land and food sovereignty show how having allies in power did not

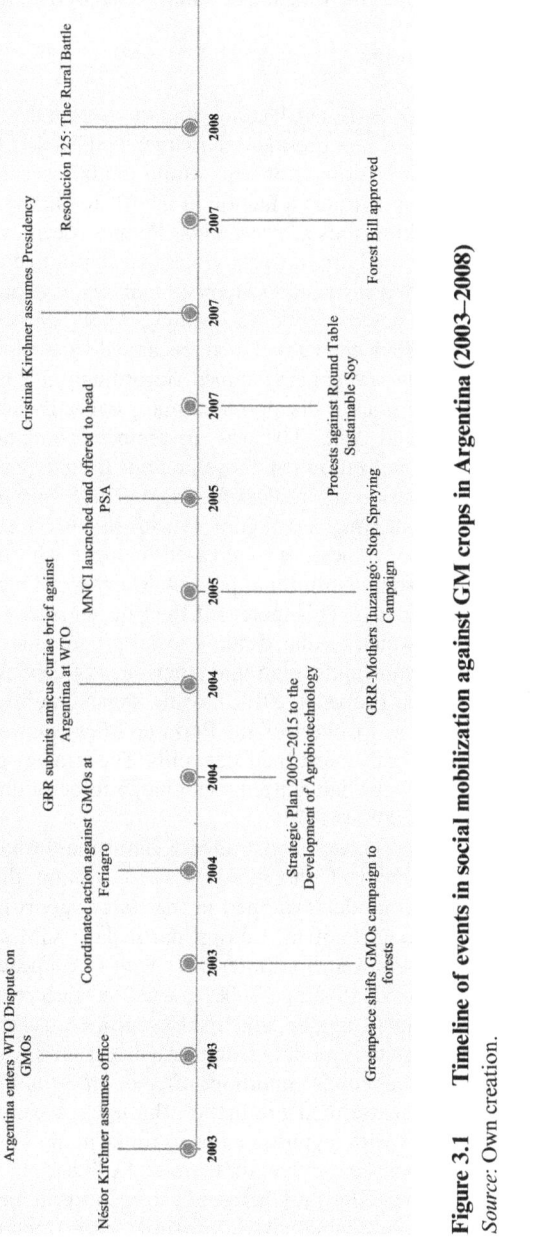

Figure 3.1   Timeline of events in social mobilization against GM crops in Argentina (2003–2008)

Source: Own creation.

translate into policy change, although some institutional spaces were opened at the margins. Meanwhile, new collective actors started to organize against the effects of the agrarian model: the sprayed peoples. All these movements observed with great interest the rift among elites that brought the agrarian model into the centre of the public agenda. Figure 3.1 illustrates the timeline of some events that marked this phase.

The beginning of kirchnerismo

In the national elections of 2003, left-leaning candidates from the PJ were elected and Néstor Kirchner became the new president. Kirchner, and his wife Cristina Férnandez Kirchner – by then a powerful senator who would in 2007 became his successor in the presidential office – inaugurated a faction in the PJ, and in the political history of Argentina, that became known as *kirchnerismo*.[1] Political decisions to advance transitional justice regarding the human rights violations committed during the military dictatorship together with a distance to financial markets have been interpreted as a rupture with the previous governments. Accordingly, it polarized the Argentinean political landscape: most parties, the PJ included, became divided along lines of support for or opposition to *kirchnerismo*, being named, accordingly, a K or anti-K faction.

The return to routine politics disappointed many collective actors who had been mobilized during 2001 and 2002. The new government combined mechanisms of integration and institutionalization (of those factions more aligned with *peronismo*) with repression and exclusion (of radical factions). The former strategy, facilitated by a faction in the PJ that sought realignment with its historical foundations, with a national-populist ideology, meant an increased dialogue with many social sectors that had been excluded from institutional politics. Many social movements expressed their support for *kirchnerismo*. This increased their dependence upon the state and, in particular, on the PJ, which mediated many social programmes, jeopardizing their possibility of adopting more confrontational repertoires of contention and thus their power as social movements (Lapegna 2013b). At the same time, historical social movements in Argentina such as Mothers of the Plaza de Mayo showed their support for the government since this advanced their demands. The strategy of repression relied on the fact that media groups had started a campaign to delegitimize *piqueteros* as a threat to order (Svampa 2008, 151–70).

There is a long debate on what constituted a change and what meant a continuity with previous governments. Concerning the issue at hand, the government gave continuity to the agrarian model, exhibited in the state support for and dependence on the production and exports of GM crops, particularly GM soy. However, there was also a change: social movements (interviews with Greenpeace Argentina, 2012; Carrasco, 2012) and analysts (Svampa 2008) started to conceptualize it in terms of a change in the development model, which shifted from a clearly neoliberal model focused on financial gains to a model relying heavily on production to export. This benefited many. The new economic conditions offered opportunities for medium- and large-sized farmers to increase their production through a technological package of genetically modified seeds with glyphosate for no-till farming. Many smaller farmers could profit by renting their lands. Overall, there was affluence in the agrarian sector, specifically in soy production. Beyond this sector, soy exports in the context of a boom in commodity prices were fundamental for a surplus in the trade balance while helping the country's economy to enter a path of positive and increasing growth rates. The government used this favourable context to renegotiate external debts, to put an end to the situation of default, and to finance social policies as compensations for poorer sectors (Giarracca and Teubal 2010; Gudynas 2008, 2012; Svampa 2012).

However, the rural poor and those living in poor neighbourhoods in the transition between urban and rural zones, usually dependent on agrarian activities (Giarracca 2003), have suffered the other side of the agrarian transformations. These include land concentration, deforestation and forced evictions of peasants and indigenous communities due to the expansion of the agrarian border. Although social movements shared this complex and articulated diagnosis of the situation, in their mobilization activities they focused on different issues due to the specificities of each organization, such as their type of membership and bases as well as their main struggles and priorities, as will be described in the next pages.

Greenpeace Argentina: shifting to forests and going global

The record harvests of GM soy in 2002 led Greenpeace Argentina to the conclusion that there was a very limited chance of their actions having any policy impact in the country. They took two decisions: they shifted attention to forests and changed the venue of the GMO battle to the global level. More than bridging two previously unrelated topics, this articulation of issues was possible due to the dynamics of agribusiness: the economic recovery and the soy boom increased land prices and the financial speculation over land, and the quest for land to produce GM soy led to an enormous reduction in the forest area in Argentina (interview with Greenpeace Argentina, 2012).

The shift to forests offered the organization many advantages. With environmental topics competing for public attention, Greenpeace sought to choose those with more chances of succeeding in terms of policy change. Data and specific events of deforestation were amplified, providing empirical credibility to claims that the model was detrimental to the environment. The action repertoire of Greenpeace, with activists dressed as jaguars climbing trees about to be felled, provided good news coverage. As the former campaigner states: 'Forests are not like GMOs, that you can't see' (interview with former campaigner Greenpeace Argentina, 2012). There was a good reaction from the general public because it was a consensual issue; it turned adversaries into allies – in particular scientists who despised Greenpeace for the campaign against GM crops – and, not of less importance, it did not face such a strong opponent like Monsanto (interview with former campaigner Greenpeace Argentina, 2012). There was the prospect of achieving outcomes in terms of legal and policy change. Indeed, in 2007, the main victory, mostly attributed to a new wave of grass-roots environmentalism (Luzzi 2012), was the issuance of the *Forest Bill*.

Greenpeace's diagnosis of the Argentinean development model, oriented towards export markets, led them to the prognostic framing of a global solution to the problem. They saw in the global commodity chain other points for activism that could be more promising for their actions. First, they targeted capital by trying to convince Argentinean exporters of the economic risks to their business due to the rejection of GM crops in Europe (Greenpeace Argentina, 2002). Then they started to send information and reports to the European offices. Greenpeace's decision to shift the focus to the external public was convergent with its organizational culture and logic, which sets as a rule the existence of real chances to achieve a positive outcome in order to go into a campaign (interview with Greenpeace Argentina, 2012).

Last but not least, the global framing of the issue also intersects the organizational culture of Greenpeace International, which seeks to define global campaigns
and differs its form of action by conducting a simultaneous action in many countries
to solve a global problem. They aim to go beyond 'not in my backyard' (NIMBY)
campaigns (interview with Greenpeace Argentina, 2012). But the contradiction that
Greenpeace Argentina did not explain is that if GMOs are part of a global campaign
from Greenpeace International, why not pursue it in their backyard as well? Indeed,
the decision of Greenpeace to drop the issue is considered by GRAIN as a big loss in
their struggle in comparison to the significant role of Greenpeace in other countries
like Brazil and Mexico (interview with GRAIN Argentina, 2013).

Instances of and obstacles to networking

While there was no sustained campaign against GM crops in Argentina, there were
moments of networking. From 2002 to 2004, two examples of joint mobilization and
networking were the Argentinean Alert Network on Transgenics (REDAST) and
Earth and Food Forum (Foro de la Tierra y la Alimentación). In 2004, they organized
a protest event during Feriagro, an important agrarian fair in Argentina. Their call for
arms was 'Argentina: land, labour and food sovereignty: for an agriculture with farmers; for healthy food for all; for the social control of natural resources'. Activists recall
that this event was unique for two reasons: 'This was one of the first times of a strong
and coordinated public activity, which reached thousands of people … and one of the
times that we managed an approximation between organizations that later, unfortunately, was very difficult' (interview with GRAIN Argentina, 2013). He explains that
the reasons for such difficulty in working together did not lie in political disagreements but rather in matters of differences in style and personal issues. Moreover, in
this time of alliance building, there was especially difficulty with the recruiting of
experts and allies in the political system. While in the former there was censorship
of dissident views, in the latter there was an overall consensus around the neoliberal
biotech food regime. Having followed the events of the GM-Free Brazil campaign, the
activist attributes their victories to these factors, which were absent in Argentina: a
network among civil society organizations, including peasants and consumers, and
support from scientists and politicians.

Going back to Feriagro, the protest banners exhibited the slogans 'food sovereignty' and 'no to monoculture of transgenic soy'. These choices signal the growing
importance of peasants' movements in Argentina as this global movement founded
the master frame of food sovereignty. Their national consolidation and increasing
involvement in the network was greatly welcomed by GRAIN:

> It is clear that the protagonists of social struggles should be the organizations of
> peasants, of indigenous peoples, and not NGOs. The political actors in social strug-
> gles should be the social movements, … who take the lead, the way; and we, to
> accompany …
>
>   (interview with GRAIN Argentina, 2013)

The activist from GRAIN recalls having started his collaboration with Via Campesina
from its very beginning in 1996. When the MNCI was established in Argentina,
GRAIN became a close collaborator.

Building bio-hegemony

Instead of adapting their supply to consumers' demands, Argentina preferred to take legal action to make Europeans accept their GM supply. In 2003, Argentina sided with the USA and Canada and entered into the largest trade dispute in the WTO. They sued the European Union (EU) for their alleged moratoria on GMOs (World Trade Organization 2006). The social movement organizations did not let their protest wait. In June 2003, Greenpeace issued a document reporting on the WTO panel and demanded that Argentina withdraw support from the USA (Greenpeace Argentina 2003). The GRR decided to act as *amicus curiae* against Argentina, siding with the EU. In 2006, the WTO panel reached the verdict, declaring the EU moratoria illegal according to the free trade regime (Motta 2013a).

Meanwhile, Argentina published its *Strategic Plan 2005–2015 to the Development of Agrobiotechnology*, in which it is stated that due to its importance demonstrated during the recovery of the crisis of 2001 agrobiotechnology is considered a strategic need for the future of Argentina and is to be promoted by the state in a cross-sectoral articulation, with collaboration of the legislative power on all levels of jurisdiction. This recommendation is based on the recognition of the fact that biotechnology is not purely a technoscientific issue that depends on innovation policy; rather, its development also depends 'on political, legal, economic variables, on external and internal negotiation' (Secretaría de Agricultura, Ganadería, Pesca y Alimentos de la Nación 2004, 5). The plan identifies labelling of GM products as the biggest threat to Argentinean competitiveness. Having shown the material power of biotechnology, the pro-GMO coalition was actively working to have the endorsement of the new government by expanding its institutional power with this policy instrument. From 2004 onwards, Argentinean authorities and traders took part in global efforts to promote public acceptance of GM soy, the most promising one being the Round Table for Sustainable Soy, a private standard that declared GM soy to be socially responsible and environmentally sustainable. Although the GRR, Greenpeace, Friends of the Earth and other organizations mobilized against the initiative, it succeeded (interview with GRR, 2012).

Launching the national peasant indigenous movement

The expression 'agriculture without farmers' calls attention to the strong process of transformation of the agrarian activity, as mentioned before. Small farmers started to question why the expansion of agrarian activities did not bring them employment opportunities and instead expelled people from the countryside. In their resistance against forced evictions, they worked in the construction of a collective identity as peasants and indigenous peasants. The creation of the MNCI in 2005 marks the consolidation of the movement in a moment of at least ten years of direct experience of local and regional efforts to resist and face the effects of the expansion of the neoliberal biotech food regime. By then, they had accumulated experience in framing grievances, building their identity, mobilizing resources, activating networks and forming coalitions with external allies. Moreover, they had expanded the action repertoire from the initial direct and immediate actions to resist evictions and deforestation (*carpas negras*) to more suggested actions by constructing alternatives and making demands on the state for public policies to support them.

The movement can be characterized as being made up of two fronts: a defensive one, which reacts to the expansion of the food regime into their territories; and an offensive one, which suggests alternatives to agribusiness. Their relative importance varies according to local realities: 'The fact that there are territorial conflicts makes it very hard to dedicate all this energy to the production. So, guaranteeing that there will be no evictions, one can concentrate on [the production]' (interview with SERCUPO, 2013). The defensive front, despite 20 years of resistance, did not suffice to diminish the threat. There is a clear diagnosis of the inequality of this fight, that is to say, the injustice felt of not having on their side politicians, the judicial system and the police as proponents of the model did: 'And we have nothing left but to fight … to resist. We have no other choice. Because we know we are in a very unfavourable situation' (interview with MOCASE-VC, 2013). Nevertheless, the movement denies the use of radical repertoires of contention based on violence. Rather, they blame agribusiness firms for bringing violence to the countryside; mobilizing police forces to arrest, inquire and intimidate activists; and establishing a process of criminalization of the MNCI members.

Here started the offensive side of MNCI, the promotion of peasant farming as the means to achieve food sovereignty, namely the master frame of Via Campesina. Just as the defensive front, the offensive front was always framed in a contentious way towards the dominant agrarian development model. In addition to their work in building identities and alternative praxis, they also demanded state policies that foster food production for internal consumption, highlighting the contrast with current policies that promote commodities destined for export. Whereas the latter bring state revenues that finance social policies, the spokesperson from MNCI would rather have policy support to maintain their cultural mode of production, to produce healthy food and live with dignity, instead of relying on social programmes:

> [I]t is not that we are complaining or we are against, no, we also have a productive project … The thing is that the successive governments never give you a chance: 'Take money and produce', [so that] the family is not a problem for the government.
>
> (interview with MOCASE-VC, 2013)

Their ability to produce food acquires the meaning of giving life dignity. The movement representatives argue that if the living conditions in the countryside ensure their farming practices and social rights, not only will rural exodus be stopped but also the migrants in the city would be able to return to the countryside. The movement also develops projects to promote food sovereignty in city suburbs (interview with SERCUPO, 2013). Recently, another issue started to receive more attention: pesticides. The MNCI learned from and supported the sprayed peoples (*pueblos fumigados*), to whom I turn next.

Sprayed peoples

Simultaneous to what happened with the peasant movements and in the urban centres, neighbourhoods living next to GM soy fields started to mobilize against pesticide spraying in their roles of victims and citizens.[2] In 2005, the GRR started to work together with a group of mothers from Ituzaingó Anexo, a locality in the province of Cordoba, with 200 cases of cancer from a population of 5,000. At the beginning of the year 2006, they motivated the GRR to organize a campaign aimed at raising awareness about pesticide use that was named Stop Spraying (*Paren de fumigar*), referring to the way pesticides were applied on a large scale by airplane or large machines. It specifically targeted the pesticide glyphosate, which is associated with the use of transgenic soy developed to be resistant to this active ingredient. However, they framed the issue beyond its technoscientific underpinnings, including a debate on the development model: the stated goal of the campaign was to stop the use of pesticides in urban areas in order to 'protect human health, environment and to launch the re-population of the countryside and food sovereignty' (Aiuto 2006).

The campaign aimed at mapping all affected localities, raising awareness by disseminating information via radio and motivating affected people to organize and mobilize themselves against fumigation. As a strategy, the campaign invited affected people from all over the country to give their testimony and share data (such as surveys of patients and analyses of water and soil) that could serve as evidence of the negative effects of the pesticides. This data was consolidated into a report and handed to national authorities with demands for corrective action.

Scientists and many social movement organizations joined the campaign, establishing the network Collective Stop Spraying (*Colectivos Paren de Fumigar*). The campaign delivered two reports (Aiuto 2006, 2009) and a book with all the documented cases (Rulli 2009). They created a manual on how to self-organize and take legal recourse against contamination from pesticides; consequently, many judicial actions followed. All these activities provided the victims with a new collective political identity; their claims were based on recognition of their rights as well as protection from the state. The campaign had many outcomes, such as mobilization, media access and changes in local legislation; however, in the first four years they had not achieved a policy change to stop glyphosate spraying (Aiuto 2009, 8).

Indeed, the decentralized competence for policies on environmental and health protection in the Argentinean federalist state led to a proliferation of local and provincial targets (Rauchecker 2015). From a case study of the municipality of San Francisco, in Córdoba, Rauchecker concludes that due to the political unwillingness of the federal government in finding a national solution to pesticide use state provinces are left with the political problem of intervening in agrarian activities, although they have no competence in agrarian policies. The solution has been to frame the issue in terms of environmental policy. The fragmentation of the solution for the local levels of an activity regulated at the national level, nevertheless, will always only be able to address the consequences and not the causes of the problem. For Arancibia (2013), the mobilization to stop pesticide spraying had many important outcomes: it generated counter-hegemonic data, spread the use of legal mobilization, and established strong alliances between experts and local communities as well as among local communities across the country, thereby building a national advocacy network on the issue. Arancibia calls it 'a broader national process of collective challenge to science base regulations of bioeconomy' (Arancibia 2013, 83). But it is not restricted to a technocratic debate on pesticides regulations; it challenges the scientific policy culture and the agrarian model behind it. Here I will tell the story of Córdoba, one among many documented in the reports from Stop Spraying; notwithstanding, this case acquired a particular importance for other movements and had national consequences, as will be shown later.

In retrospect, the ex-undersecretary of health in Córdoba (interview with Ávila, 2012) identified that from 2000 onwards there had been many denunciations from small villages and neighbourhoods about spontaneous abortions and malformations. However, people usually treated such events as personal problems that should not be made public, trying to find the cause of the problem in their own behaviour. The emergence of social mobilization in Ituzaingó Anexo[3] can be attributed to a shift in grievance interpretation: from self-blame to external targets of blame.

In the year 2000, a mother, Sofia Gatica (interview with Mothers of Itunzaingó, 2012), after having suffered the death of a newborn child, started to ask herself why many of her neighbours were ill. She heard of other cases from other mothers in the neighbourhood and they started to share their opinions that when the airplanes sprayed pesticides over the soy fields that surround their houses many suffered with immediate reactions to their skin and their breathing as well as from headaches. In this way, they associated their health problems with the practice of spraying pesticides. They became known as the Mothers of Ituzaingó, defining their struggle as defending the lives of their children.

Here an excursus on gender and the role of women in Argentinean rights movements is necessary. The use of gender to mobilize for human rights in the country is known worldwide due to the organization of the Mothers of the Plaza de Mayo. There is no simple proposition on how to relate women and human rights; as points of reference, one can historicize it as a construction and reveal its contradictions (Jelin 1996). On the one hand, their gendered mobilization can be criticized as conservative since they draw on the traditional role of mothers as moral protectors and representatives of the family, and then extend this to protest against human rights violations that put their family under threat. On the other hand, their use of the identities as mothers is strategic as it allows them to bring issues to the public sphere that would otherwise be much harder to make visible and politicize (such as the human rights violations during the dictatorship and now associated with the agrarian model). Although the Mothers of Ituzaingó might not have consciously and strategically drawn on this 'historical frame of women's political participation' – meaning that women are representatives of the family, of the public morality and of nation as a family (Bonner 2007) – it was part of Argentinean political culture, therefore it resonated among the people. The difference to their historical forerunners lies in the class dimension, as the Mothers of the Plaza de Mayo are from the middle class whereas the Mothers of Ituzaingó belong to a poor neighbourhood in the rurban zone.

In order to make their claims credible, the Mothers of Ituzaingó made their own epidemiological map. They mobilized their main resource: their network in the neighbourhood, which gave them access to all families for collecting data on their health situation. They found that the incidence of cancer and abortions in their neighbourhood by far exceeded average rates. This resulted in an innovative contentious tool to fight GM crops – an action repertoire by no means related to traditional roles of women – that is to say, the construction of 'counter-hegemonic epidemiological data' (Arancibia 2013). The Ministry of Health from the province did not react to their data. A protest followed with the slogan, 'Help us, we have cancer', blocking the roads and reaching media attention. Local authorities heard their demands but did not recognize that the pesticides were dangerous to their health. However, they sent experts that confirmed the health situation of a contaminated community and issued three municipal ordinances in 2002 that declared the neighbourhood in a state of emergency, prohibited air spraying of pesticides in the city of Cordoba and prohibited land fumigation in Ituzaingó. In the same year, they brought the case to court; however, they faced enormous obstacles in advancing the legal action.

Witnessing the continuous practice of pesticide spraying over their homes, the mothers kept up their protests such as roadblocks in which the ill people from Ituzaingó took part. They received death threats and realized that the police, instead of enforcing the ordinances, was on the side of soy producers. Thus they decided they had to mobilize more scientific knowledge, draw allies to their side and bring their demands to the judicial system. They learned a lot from searching on the Internet. In 2005, they went to Buenos Aires to demand national authorities find a solution to their plight. While these demands did not succeed, their presence in the national capital awoke the solidarity of other organizations, like the GRR. As their message spread, with the aid of media coverage, many (national and foreign) social movement organizations contacted them, offering support in varied ways.

Gatica recalls that it was through the support provided by the Network for a GM-Free Latin America (RALLT), represented by Elizabeth Bravo, that they learned what a transgenic seed was and that it was developed to be resistant to pesticides, thereby leading to the use of more pesticides. So, they understood what was happening because transgenic crops surrounded their houses. By then, they had already developed a clear diagnosis of the social and environmental injustice of their situation as a contaminated community. They framed it in terms of discrimination and unequal distribution of environmental burden, 'as the price that we have to pay for a supposedly progress that benefits a few' (Grupo de Madres de Córdova 2005). They stated that their fight was to make visible what the country has been trying to conceal: the victims of its record harvests and of the adoption of the transgenic agro-exporter model, which they characterized as the cause of their suffering.

The Mothers of Ituzaingó did not stop short of attributing responsibilities for the health and environmental contamination of villages nearby soy fields: they considered it to be a crime marked by the state's omission in not enforcing environmental laws and thus showing connivance with transnational corporations. In sum, this group of neighbour mothers who started to act in the defence of the health of their families developed a detailed action repertoire, including the production of counter-hegemonic data, the formulation of law projects, legal mobilization, as well as the establishment of an increasingly national and transnational solidarity network. Finally, they motivated an ally in the Ministry of Health from the municipality of Cordoba, Medardo Ávila, who required the Public Prosecutor's Office to take judicial action against three men who violated the ordinances. The case reached the courts in 2008. This was also the year of another conflict; however, in this one the powerful representatives of agrarian producers were present and went to the streets to protest, as will be described next.

'The rural battle': a window of opportunity?

In 2008, the soy model reached the national public agenda. Since 2002, the volume of taxes collected on agrarian exports increased following the rise of international prices; as these were highly favourable, the objections from agrarian producers had no major impact. On 11 March 2008, the recently elected president, Cristina Fernández Kirchner, issued Resolution 125, which changed the system of export taxes: the fixed rates would be substituted for a mobile percentage, increasing taxes for soybeans (from 35 per cent to 41.1 per cent) due to the increase in international prices and decreasing it for the other grains. Given the record in international prices that year and farmers' expectations of record profits, the resolution generated a strong protest.

For the first time, members of the traditional landed elite, represented by the SRA, adopted disruptive direct action such as roadblocks, which was not part of their action repertoire since they had always had access to power. Indeed, one of their main demands was the opening of institutional channels to start a dialogue with government, as the measure had not been previously an object of consultations with the sector. An unprecedented coalition of the most prominent representatives of the rural sector was formed, the Mesa de Enlace (Liaison Committee), composed of the organizations SRA, FAA, CRA and CONINAGRO. For the first time, the smaller partners of the agrarian sector joined the big ones, a sign that the biotech food regime was also beneficial to many medium and smaller farmers, which also had consequences in their political alliances.

All societal actors not directly involved in the conflict had to take up positions, including economic actors and associations, political parties, media groups, personalities and intellectuals, union leaders and other representatives from civil society. Also the urban middle class took to the streets banging pots.[4] The polarization of the political landscape between the government and the countryside reactivated the recurrent process of creating binary oppositions in the Argentinean political history, which tend to exacerbate conflict to an extent that a democratic plurality of arguments and positions is reduced to two apparently monolithic blocks (Svampa 2008, 231–233). Instead, their strength was disputed on the streets by the size of the protests supporting each side and, finally, in parliament by votes.[5]

Lasting more than four months, the 'rural conflict' or 'the rural battle', as it became known, was the central event of the year (Aronskind and Vommaro 2010; Giarracca and Teubal 2010). In dispute was the share of rents between agrarian actors and the government. Although the conflict was all about the percentages of taxes, leaving the issue of GM soy monocropping untouched, it paved the way for activists – *piqueteros*, trade unions, the CTA, intellectuals and individuals forming part of an independent left – to discuss other issues related to the model. Their argumentation highlighted the social uses of production according to the needs of the public. They defended the export taxes but proposed specific destinations for them, such as to lower taxes over food products and to increase wages and social security. They also demanded a stronger role of the state in regulating agrarian polices as well as in the use of natural resources ('Otro Camino Para Superar La Crisis' 2008).

Notwithstanding the 'dislike' on the part of the agrarian elite, both old and new, regarding the government of Kirchner – for instance, the president did not go to their annual fair to show their prestige in the society – they were consistently favoured by state policies. Beyond the arm wrestling between the government and the Mesa de Enlace, the role of soy crops in Argentinean economic policy was never questioned. Rather the contrary seems true: the government continued promoting soy crops, as the Federal and Participative Strategic Planning of Agrifood and Agroindustry (PEA) confirmed later, foreseeing the increase in grains production (Ministerio de Agricultura, Ganadería y Pesca 2010). This plan, formulated together with the agrarian sector, showed how the model was consolidated institutionally in Argentina, as if it was a state policy that no government, with its rhetoric of progressiveness and anti-neoliberal notwithstanding, would dare to change. The conflict of 2008 was, in the end, a mere disagreement inside the consensus for bio-hegemony.

Svampa (2008, 29–30) summarizes the meaning of the conflict over Resolution 125. The results are ambiguous: on the one hand, the conflict provided a window of opportunity to discuss the model. On the other hand, it became clear that the government maintained a very linear and productive vision of development. In short, the conflict showed that there would be no political opportunities to challenge it due to the underlying convergence between the government and agribusiness.

A rising cycle of contention over 'the model' (2009–2013)

After the rural conflict it became clear that the government would only deepen the agrarian policy anchored in commodity export. Conversely, there was also a growing mobilization against its effects, a process that will be described next in four steps. This also intensified land conflicts and violence in the countryside, and the peasant movement responded with increasing mobilization and diversification of action repertoires, alliances and demands. The increase in pesticide use became the new entry point for social movements to dispute the agrarian model. The contradictions, however, brought by the fact that the current government supports and is dependent on agribusiness – and still is perceived as the better option by social movements, always comparing it to the neoliberal decade of *menemismo* – places movements in a very complex context for their mobilization efforts. On the one hand, there are prospects for a new cycle of contention. On the other, activists disagree on where the solutions to overcome the development model based on neo-extractive industries and agrobiotechnology lie: global or local solutions, national or regional political agency? The chapter finishes with this open debate. Figure 3.2 summarizes the last events in Argentina on this contestation.

Peasants: violence, law-making and reaching new spaces

In the process of expansion of the agrarian border, due to forced evictions, land conflicts and conflicts over deforestation, violence and murder increased, in particular in northern Argentina, where there is a historical record of human rights violations. These are often not investigated and left with impunity. Building on the authoritarian tradition in those regions, which were also the places where the agrarian border was expanding, the hegemony of the neoliberal biotech food regime reacted to protesters with violence. Activists were confronted with constant threats, fearing for their lives, without having the police or the judicial system on their side. The criminalization of social movements and protest was institutionalized with the approval in 2011 of the *Anti-terrorist Law* (GRAIN 2013a, 4).

Violence in the countryside started to be documented by the Ombudsman's Office (*Defensoria del Pueblo de la Nación*), which focuses on indigenous communities, by networks of academics and activists (UPC- UNESCO *et al.* 2009), and by NGOs. Since 2009, the NGO Red Agroforestal Chaco Argentina (REDAF 2014) conducts annual surveys on land and environmental conflicts in Chaco. The organization attributes most conflicts to the expansion of GM soy. Among the victims, they name indigenous communities and small farmers; however, above

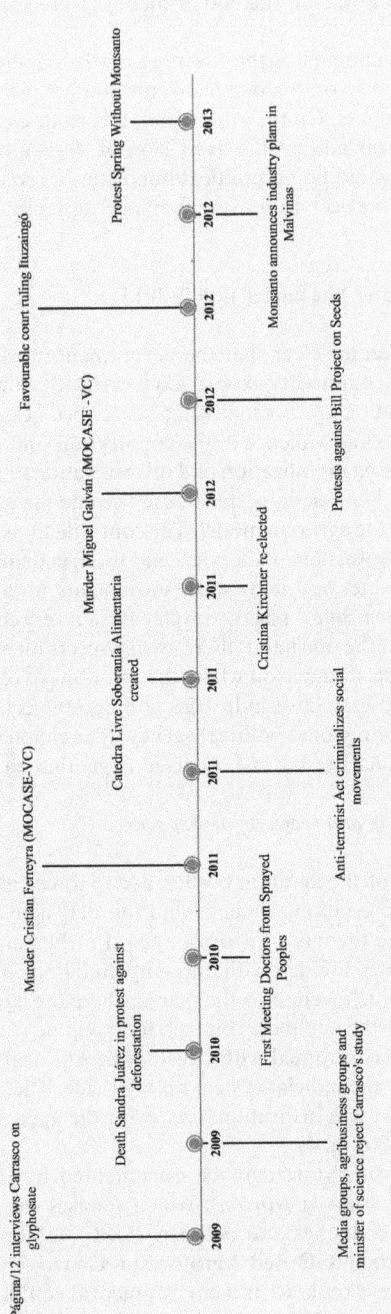

Figure 3.2 Timeline of events in social mobilization against GM crops in Argentina (2009–2013)
Source: Own creation.

all the emphasis is placed on the expansion of the soy model as part of a clash of cultures and identities in regards to land use (Aranda 2010). From 2010 to 2013, three activists from the peasant movement MOCASE were murdered in Santiago del Estero: Sandra Ely Juárez, Cristian Ferreyra and Miguel Galván.[6] In northern Argentina in general, the cases of murder, violence and death threats have increased.

Although *kirchnerismo* advanced the agenda of human rights organizations regarding the crimes committed during the dictatorship, it was often complicit with human rights violations that occurred in the provinces. Lapegna (2013a) explains that violence against peasant and indigenous movements is not punished due to the constraints of realpolitik. *Kirchnerismo* relied on political support from provincial governors, who control party votes in the national congress. In exchange, the government does not recognize, condemn or punish human rights violations in these places, with the complicity of the mass media.

Without discursive opportunities in the media, the MNCI reacted to these circumstances with protests and institutionalized their demands in a law project to stop violence in the countryside. It was named after the member of the organization murdered in 2011, Cristian Ferreyra. Intended as a general law on peasant rights, its scope had to be restricted to the issue of forced evictions. Interpreting their situation as being defensive rather than in the condition to make an offensive, that is to say, to make more demands on the state, they considered it would be already a good outcome if they passed a law that stopped forced evictions. The movement interpreted that such constraints were due to the fact that the issue of land rights and land reform in the countryside had not reached the public sphere as a political problem, and peasants were not recognized as a political subject: 'Argentina has been disregarding the existence of peasantry for two decades, in the academy, in the political sector' (interview with SERCUPO, 2013).

However, the MNCI achieved a concession from the national government in 2008: the launching of the Under-Secretary of Rural Development and Family Farming (SSDRAF) at the Secretary of Agriculture, which was transformed into the Ministry of Agriculture. The SSDRAF meant, according to Lapegna (2015), an act of 'institutional recognition' to the peasant movement in exchange of political support to the national government, in particular after the rural battle. The movement was aware of the limitations of the new undersecretary, occupying a small area of a ministry dominated by agribusiness, and being third in line in the hierarchy. Nevertheless, for the movement he represented an ally in power and a positive prospect of opening the political space for their demands on peasants' rights, even though they had a realist understanding that this would not mean a structural change in the agrarian policy (interview with SERCUPO, 2013).

Meanwhile, in 2012 the Ministry of Agriculture announced a bill project to modify the *Seed Bill*, incorporating demands from biotech firms and big agrarian producers. Its main sponsors were the Argentine Seed Association (ASA) and Monsanto, which had been demanding this law for over a decade. Monsanto threatened not to launch its new GM soy in Argentina if a new law was not in force. It was designed to adapt the Argentinean legal frameworks to the neoliberal free trade regime, in particular in relation to intellectual property rights and seeds. Its aims were to remove obstacles against a further commodification of seeds and genetic material by guaranteeing intellectual property rights as well as the collection of royalties while, at the same time, revoking farmers' use, which means farmers' rights to multiply and exchange seeds for non-commercial purposes.

The peasant movement was mobilized against the project (MOCASE-VC, 2012). In addition to the old partners (GRAIN 2013b), the MNCI counted on the support of critical spaces in scientific and educational arenas. In the humanities, there had long been study groups on the agrarian issues in Argentina, which positioned themselves politically against the neoliberal food regime. In the agronomy faculties, however, the study of the issues of relevance to peasants implied a challenge to the dominant curriculum, focused on agribusiness since the 1990s. There were many contracts between university and firms specialized in agrarian inputs (pesticides, seeds, fertilizers, machines), through which the latter financed research projects, events and even public and private university buildings (interviews with SERCUPO, 2013; Luque, 2012). Alternative thinking started to emerge from the fieldwork of some university students and professors, which culminated in 2011 in the constitution of the Free Chair on Food Sovereignty (Cátedra Libre de Soberanía Alimentaria) at the Faculty of Agronomy at the University of Buenos Aires, which supported the protest against the *Seed Bill* (Cátedra Libre de Soberanía Alimentaria 2013). Although Via Campesina required for membership identification as peasant organizations, the participation of university students, professors and researchers in the MNCI grew steadily. In sum, the peasant movement could not abandon its defensive front as violence indeed increased in the last years; however, the movement expanded its action repertoire to include more institutional means such as formulating law projects, making demands for the establishment of public policies and institutions, and finding access to the political and educational systems.

Pesticides: a new coalition?

The Mothers of Ituzaingó and the Collective Stop Spraying have signalled to other movements that the pesticides issue could be a better entry to their dispute over the agrarian model. To some extent, they might be considered as early risers in a protest cycle that is in the making. Physicians from small villages, who had been calling attention to the health situation of areas surrounded by soy fields for years, and scientists joined efforts in 2010 when the University of Cordoba organized the first meeting of Doctors from Sprayed Peoples (Médicos de los Pueblos Fumigados). This signalled dissidence from the prevailing official position of the medical school by recognizing that pesticides are toxic. Dissident scientists have indeed been very marginal in the Argentinean history of resistance to GM crops. But a prominent scientist advanced his expertise for the struggle: the story of Andres Carrasco exposed the long active censorship reigning in public research institutes. He testified in the court for the Mothers of Ituzaingó. Their case, having reached the courts, drew attention and solidarity to their fight and intensified protests against the agrarian model. Some organizations saw a new and more promising chance to fight, while others criticized the shift to the pesticide issue, showing a 'framing dispute' (Benford 1993). In any case, some events have already served to make visible structural conditions for mobilization. These events will be told in three steps.

The controversy around the researcher Carrasco: science vs politics

A prominent researcher from the National Council for Scientific and Technical Research (CONICET) and head of the Laboratory of Molecular Embriology from the University of Buenos Aires, Andres Carrasco became interested in the issue of pesticides from hearing about the Mothers of Ituzaingó. From his travels to the hinterland and reports from physicians on the cases of cancer and malformation, he suspected that it would also happen elsewhere. He played a crucial role in Argentina by conducting research on health risks of pesticides associated with GM seeds (Paganelli *et al.* 2012), contributing to the very few efforts in this direction internationally (Mesnage *et al.* 2013; Séralini *et al.* 2012). His involvement in the issue, however, did not occur without awakening counter-reactions from the coalition that maintained the dominant agrarian model. In the beginning of 2009, Carrasco's role was pivotal in a political dispute on the health effects of glyphosate.

In April 2009, a fierce public debate ensued after the newspaper *Página/12* published an interview with Carrasco (Aranda 2009). This window of opportunity was not contingent. Rather, Carrasco's decision to publish his results in the mass media was based on the fact that there was a journalist who would be ready to hear his story and would make all efforts to have it published: Darío Aranda. Aranda had a long history of writing on indigenous and peasant movements and was recognized as a distinguished critical voice on the agrarian model. Although he had been working for *Página/12* for several years, he was never employed, which left him in a precarious situation, never sure of whether his pieces would get published (interview with Aranda, 2012). Aranda managed to still get the go ahead of the *Página/12* editorial team, which is known for its pro-*kirchnerismo* position (interview with Carrasco, 2012).

Based on Carrasco's study, the Argentinean Association of Environmentalist Lawyers (AADEAA)[7] took judicial action in the Supreme Court to suspend the pesticide and the minister of defence prohibited the use of glyphosate in the lands under its supervision. The AAPRESID, the CASAFE and the Argentinean Soy Chain Association (ACSOJA) reacted vehemently with the aid of media groups associated with agribusiness, in particular *La Nación* (Motta and Alasino 2013). The minister of science and technology declared that the study was not an official study from CONICET.

Carrasco realized that the problem did not lie in his work but in how he communicated it:

> when I present my data, I frame it in ideological and political terms: I end up criticizing the model. This model leads us to this. We are speaking about glyphosate for a simple reason: We have 20 million hectares and have used ... 100 million kilos of glyphosate per year.
>
> (interview with Carrasco, 2012)

He highlights the uncertainties over the implications of genetic engineering over evolution and the ethical commitment of science of acting with responsibility. This is why to him it is necessary to discuss the GM technology because it lies at the core of the agrarian model: 'The profitability is based on the advantages of the technological dispositive. In other words, the GM seeds are not a casual invention, they are a desired invention ... If one takes out GM seeds, glyphosate is useless' (interview with Carrasco, 2012). It became clear to him that had he just published scientific results in a scientific journal, he would increase his scientific capital. The problem is that he crossed the lines: aware of the ethical borders of science and the implications of his findings in the wider society, he tried to influence a public debate and went to the public sphere of mass media to speak out.

Although officially created in January 2009 to investigate the health effects of pesticides under the coordination of the Ministry of Health, the National Commission on Pesticides was actually set in motion by the minister of science and technology, Lino Barañao, following the Carrasco episode. Famous as a first-line supporter of *kirchnerismo* and of the agribusiness model, the minister reacted to Carrasco's study by installing a committee to make a literature review on glyphosate. The report does not mention Carrasco's study at all, although it is full of indirect references to it. It specifically states that although glyphosate can change the DNA structures in animal cells, 'it is improbable to find the concentrations of effect in human biological environments, what leads to the assessment that there would be no significant risk to human health' (Comisión Nacional de Investigación Sobre Agroquímicos and CONICET 2009).[8] In other words, the commission expressed a risk culture that disregards evidence of risks and the limits of science, and neglects the precautionary principle.

While this became the official top-down position from CONICET and the Ministry of Science, other researchers from CONICET published a petition criticizing the reactions to Carrasco's studies and the way in which universities and scientists were paid by transnational corporations to legitimize the hegemonic discourse on the development model (Voces de Alerta 2009). Observing that the national public sphere in Argentina was closed to him, Carrasco moved between scales: going local and going global. Locally, he became an important ally of sprayed communities and Collective Stop Spraying. Globally, his work called attention to the global anti-GMO movement. The German Green Party invited him to go to Brussels in 2010 and speak to the European Commission, and to Berlin in 2011.[9]

Carrasco is aware of the GRR's criticism of him for highlighting the issue of pesticides, making the debate lose its focus. However, for Carrasco, it is legitimate that directly affected people do it, since it is their fight. He adds: 'it is also right that the debate about effects of chemicals catalysed the big national debate, at least about chemicals' (interview with Carrasco, 2012). But he warns against the danger of playing the game that the adversary wants: 'Monsanto does not want a discussion about the model; they prefer a discussion about agrochemicals ... while we do it, we do not discuss the other, the politics' (interview with Carrasco, 2012). Thus, for him the way forward must include a debate on the wider framework of the issue. But his story is also an example of the threats faced by those who speak about the model in Argentina due to the high stakes involved for powerful economic actors and their coalition in the national government.[10]

Ituzangó in the courts

The legal case of the Mothers of Ituzaingó was judged in 2012. This case presents an important departure from previous judicial rulings on pesticide use because the misconduct of three individuals who sprayed pesticides over GM soy fields was penalized as a criminal act of environmental contamination and not as an administrative failure. Although these men actually carried the whole burden of responsibility, the public prosecutors elaborated upon the underlying causes – the agrarian policy – and made recommendations for legal and policy changes that challenge not only the Argentinean regulations but also the global state of the art on pesticides.

Challenging the often invoked argument that the use of pesticides is a safe and legal activity guaranteed by the National Service for Agricultural Health and Quality (SENASA), the prosecutors drew on the precautionary principle, foreseen in the *Environmental Bill* (Law 2565). They argued that the authorization of pesticides cannot be invoked when there is contamination and there is evidence of damage to human health and the environment. The strategy was to contrast this with the bureaucratic procedure of authorizing pesticides – based on laboratory data on animals – and the actual use of the product after approval. More than an issue of bad agricultural practices,[11] the public prosecutor called attention to the dimensions of pesticide use in Argentina. The argument was that although pesticides like glyphosate are classified by SENASA in the lowest classes of toxicity, any pesticide may cause damage depending on the degree of exposure. Using data from CASAFE, the public prosecutor mentioned that in Argentina 300 million litres of pesticides were used in 2010 on 22 million hectares, affecting 12 million people living in those areas. As evidence of damage to human health, they referred to the report from Doctors from Sprayed People and to a health survey conducted in 2010 in Ituzaingó, emphasizing that 33 per cent of the causes of death were attributed to cancer, much higher than the national average, as well as the high number of spontaneous abortions.

In addition to penal sanctions and community services for the three accused men, the public prosecutors recommended that the Ministry of Health should send a *National Bill on Pesticides* to the Congress that prohibits aerial spraying and sets limits on terrestrial spraying. SENASA was to reclassify all toxicological products by taking into account chronic intoxication, to prohibit highly toxic products and to ensure that toxicological studies are conducted by public laboratories or universities. The local Ministry of Health in Cordoba was ordered to prohibit air and terrestrial sprayings within 1,000 meters of villages and to initiate campaigns to enforce the law and to educate producers in pesticide applications.

Activists interpreted the judicial rulings in varied ways. The former undersecretary of health (interview with Ávila, 2012) emphasized the symbolic change achieved by this case. The cultural and historical legitimacy of the farmer in Argentina, espousing a social identity of hard worker and morally charged with a strong symbolism, was an obstacle to the debate on the negative effects of the agricultural activity. Sofia Gatica (interview with Mothers of Itunzaingó, 2012), although satisfied that justice was done in the particular case at hand, was well aware that the three accused people had only a very limited responsibility. For her, the ultimate responsibility lay with the government and the multinationals. Many in the Collective Stop Spraying network found that settling the issue in the public agenda could already be considered a victory (interview with Semillas del Sur, 2012; Luque, 2012). However, they believed that the judiciary could not solve it alone and a political solution was needed. This was constrained both by the difficulty in mobilizing people who receive cash transfers from the state and by the interests of the state in the taxes over soy. Their conclusion was that they must continue fighting for the support of the public opinion.

Indeed, their mobilization potential has only increased. On 19 September 2012, a month after the court ruling, they organized a protest where 10,000 people participated. The Mothers of Ituzaingó kept leading most protests. More than being the early risers, GRAIN believes that Mothers of Ituzaingó can be a factor to bring cohesion in a new coalition in Argentina around pesticides, not least due to the strong symbolism of mothers in the Argentinean politics. In fact, representatives of the Mothers of the Plaza de Mayo have shown their support for the Mothers of Ituzaingó during the legal proceedings (*La Voz Del Interior* 2012). Many other organizations have taken up the pesticide issue recently.

Following the track of the early risers: some contradictions

The MNCI saw pesticides as an issue with already some positive outcomes and with some hope for future victories: The perception of success from the Collective Stop Spraying network, the resonance of the issue in the countryside and the cities alike and the involvement in the continental campaign from Via Campesina, launched in 2010. Following that, the MNCI officially announced the Argentinean campaign in 2012.

However, not much has actually been done since the official launching of the campaign. A main obstacle for the MNCI has been the ambiguity in their discourse regarding the allocation of responsibilities in the Argentinean polarized political landscape. Given the complexities of political alignments, the blame game is not easy to follow. At the national level, the MNCI have supported the national government and have allied with *kirchnerismo*. In the national presidential elections of October 2011, they supported the re-election of Cristina Kirchner. This hindered their mobilization efforts and autonomy as they did not clearly attribute responsibility to the president for the current problems and in delivering solutions. At the local and provincial level, the situation varied. When not supporting local governments, this independence allowed them to shift blame from the national to the regional and local levels.

The following quotation referring to the establishment of an industrial plant by Monsanto at Malvinas Argentinas shows the ambiguity of MOCASE-VC regarding political responsibilities: 'Maybe she [the president] has no idea of what they are signing behind her back, say, because many things go through this way, you know, many things happen provincially; and nationally, but through other means, do you understand?' (interview with MCC, 2013). The blame on multinationals was recurrent in their speech. Not only is this in tune with their framing activities conducted in Via Campesina International, which frames the enemy in a global fashion, but this is also convenient due to the political alignments of the MNCI in Argentina. The MNCI were keen to highlight the divisions inside the government. To this end, they named the regional Ministry of Production or specific sectors of the political establishment (not all) that supported the multinationals (interview with MCC, 2013). They perceived that the political opportunity structure was negative towards their overall struggle:

> The current government is very strongly committed to the GMO model as a form of obtaining revenues. And with the current scheme of 35 per cent export taxes, well, it is an important entry of foreign currency with which there is social policies, an important part of the current policy.
>
> (interview with MCC, 2013)

In addition to the political alignments to sustain the model, the MNCI identified constraints for their action in the media structures and the scientific institutions in Argentina, given the economic stakes that the major media groups and research institutes had in agribusiness (interview with MCC, 2013). Nevertheless, their perception of the opportunities for mobilizing remained optimistic. Such perception can also be explained by an overall rise in contention regarding socio-environmental issues in Argentina, the object of the next section.

A new cycle of protest

Reading such structural threats for their action in Argentina, the MNCI emphasized the role of civil society and social movements in bringing societal change. It is interesting to note that a realistic reading of the structural constraints only reinforced their belief in the power of collective action, not the contrary as expected (Gamson 1992), not least because the representatives of the MNCI had always stressed that due to the structural inequalities in their access to politics they had no other choice than to fight. The hopes of the MNCI and many other movements and activists were thus placed in a new cycle of contention in Latin America. Some analysts have indeed identified an epochal change (Svampa 2008) in social mobilization in Latin America. It is to the movements' perception and interpretation of a rising cycle of protests that this section turns. They might disagree regarding the main demands at stake, but they shared the perception of a protest cycle on the rise, bringing new opportunities for building coalitions between movements fighting different issues – such as mining and the agrarian model – and creating a new momentum for winning the battle against GM crops.

Apart from Greenpeace's short-lived actions on labelling GM products, which relied on the fear of an urban population vis-à-vis a new technology, all movement actions touching on the issue of agrobiotechnology focused on the people directly affected by the agrarian activity, interpreted in an articulated master frame of a development model. Environmental and health problems were thus the immediate consequences of that, but not the ultimate causes of the grievances. The MNCI identified signs of a new type of politics – a change – drawing on the experience to the previous cycle in 2001:

> And it seems to me that there is a new politics, the social movements have a lot to do with this change. We can say it starts from 2001, when society was already very troubled and well, there was no other exit than to go out to the streets.
>
> (interview with MOCASE-VC, 2013)

Greenpeace also distinguished the current mobilizations from previous ones, which were more focused on employment, whereas linkages between social and environmental issues characterized the new wave. The explanation for this change lies in a combination of changes in media coverage, long-term societal change achieved by environmental education, and the accumulation of resources by environmental organizations in the country. The processes of mobilization resembled those observed in the neighbourhoods affected by spraying or the peasant and indigenous communities threatened by land evictions: they started at the very local level as a response to the direct threat, that is to say, first it was a NIMBY reaction. These mobilizations started to be identified by their form or organization (*asambleas*). Then local conflicts learned from one another, developed networks and they started to get involved in other environmental issues.

Whereas for Greenpeace the environmental consciousness was a result of long-term process and was central to their organization,[12] Carrasco interpreted the protests around environmental issues as the only desperate reaction to immediate suffering. To him, the underlying problem was political and economic, not environmental. In addition, he saw fewer prospects of building on the victories achieved in the resistance to mining to mobilize against the agrarian model. To him, there were significant differences between targeting one multinational firm with a mining project in one location and criticizing thousands of Argentinean farmers, as the agrarian activity has an economic importance for households and small cities and villages, and is a source of social upward mobility in Argentina. Culturally, farming has a strong centrality in the Argentinean self-understanding of their history and culture, so making it an adversary makes it much harder to find resonance with the public, whereas mining is a negatively framed economic activity, often associated with colonial powers (interview with Carrasco, 2012).

All these obstacles led to the conclusion, according to him, that the problems arising from agrobiotechnology would have to become worse in order to be visualized. He did not see resistance disseminating in society and integrating other actors who were not directly affected. He was very impressed by what he saw travelling in the Argentinean countryside, and remembered an activist telling him in a kind of 'agonistic scream': 'Do you know, Doctor, they are changing our lives ... they are killing us not with toxics, but they are destroying our lives' (interview with Carrasco, 2012). Carrasco's prognosis was murky: the model would depopulate the territories and pursue commodification of the lands in Argentina to the limits of an extractive model, 'because, as long as there are people, there is resistance, there is a nucleus of dialogue' (interview with Carrasco, 2012).

GRAIN was much more optimistic. In contrast to the past difficulties in working together, the organizations started coordinating joint activities such as the protest day 'Spring without Monsanto', on 19 September 2013, against the installation of the Monsanto industrial plant in Malvinas Argentinas, Córdoba. In addition, for GRAIN there were already partial victories, such as the regulation of pesticide spraying in many localities. They maintained the focus and the hope on the policy reversal of GM technology, albeit acknowledging the power of their opponents:

> [I]n spite of all that, the way is to question GMOs, because its uselessness in feeding the world and producing better or more food was demonstrated, the risks as global experiment were demonstrated. There are some ideological battles that, much beyond the praxis, are being gained.
>
> (interview with GRAIN Argentina, 2013)

In sum, the perception of a rising cycle of contention around the development model articulated environmental and social issues, both for economy and politics. Health and environment damage and risks from technologies became more salient due a long process of building an environmental consciousness in the country. At the same time, the advancement of the extractive and agrarian industries exacerbated conflicts; there were mobilizations not only around direct health and environmental suffering but also for claiming land rights and the recognition of cultural identities. All these issues have been discussed by an increasing number of assemblies and meetings; new networks were established, and some saw that in the ideological battle there were signs that GM crops were defeated, even if the state policy lagged behind due to powerful economic interests. There was an ongoing framing dispute among organizations in Argentina regarding the locus of solution to the problems identified with the Argentinean developmental model, as will be described in the next and last section.

Global vs local, national states or Latin America?

Activists were also working to identify the better scale on which to find solutions to their struggles. On the one hand, all interviewees identified the global dimension of the agrifood system in which the Argentinean agrarian model is inserted. Consequently, they named the difficulties in finding isolated or country-specific solutions. At the same time, some activists highlighted the abstractness of global responses when it comes to local realities. On the other hand, some activists identified the national state as the main addressee of their claims, even though they recognized the responsibility of global economic forces and actors. To further complicate things, most activists saw no political opportunities in Argentina to advance their demands nationally. Many thus believed that the solution lay at the local level, from which struggles can scale up. These varied positions will be briefly illustrated in this section.

For more than 20 years, GRAIN has worked to influence international treaties at global venues, relying on its professional capacity to generate, analyse and disseminate information on both the grass-roots level as well as the legal and political institutional changes at national and global levels. It could enumerate many achievements in global treaties. 'But we saw that people's lives have not been transformed at all' (interview with GRAIN Argentina, 2013). GRAIN highlighted the difficulty of making the global victories reach the local level and turned its focus more and more to this level of activity.

In addition to the dichotomy global vs local, some activists referred to the alternative between global and national levels. This is the case for Greenpeace. The rural conflict in 2008 only reinforced Greenpeace's perception that there was no room to fight GMOs in Argentina. Greenpeace saw the solution at the global level, though not necessarily in international legal agreements. Rather, Greenpeace Argentina was constantly referring to global market forces: 'Many issues can be resolved. Others won't be solved only by Argentina; they will be solved by the international context ... Not because people achieved that, but because there is less pressure, to produce soy is not business anymore' (interview with Greenpeace Argentina, 2012).

The shift of their actions on GMOs to the global problem might be more efficient and more congruent with the operational logics of Greenpeace, but many other activists in Argentina, without neglecting the global dimension of issue, challenged this interpretation and strategy for a number of reasons. They claimed that treating it as a market issue and trying to influence foreign consumers and foreign governments might not turn out to be a good strategy. In Europe, Carrasco encountered a cynical reaction to his appeals to raise awareness of the local effects of European grain imports. In hindsight, he acknowledged that there was not much reason for hope as Europeans basically blamed the Argentinean government for its decision to plant so many hectares of GM crops.

Carrasco was aware of the risk of obfuscating the political decision in Argentina, and thus called attention to its contingency that it could always have been different. The responsibility of multinationals notwithstanding, the addressee must be the national state, where the political decision on the development model and use of natural resources lies: 'I am not debating with Monsanto. The only valid interlocutor is those in charge of political decision, who can change the model or start to think of alternative models, and this is the national state' (interview with Carrasco, 2012). He compared Argentina to many Latin American countries to make the case that it was not a mechanical technological diffusion but a result of a political decision; there is always a moment of decision: 'That multinationals can impose a technological model as the GMOs is right to a certain extent, but to another extent no. But the answer could had been no' (interview with Carrasco, 2012). Evoking a general meaning of sovereignty, he recalled the publicity piece 'United Republic of Soy', a name coined by Syngenta to refer to the southern cone of Latin America, where soybean production knows no borders. In other words, an advertisement that made a mockery of political decisions:

> The same happens in Paraguay, Brazil, Uruguay, the media luna in Bolivia: agrarian areas of more than 50 million hectares that have ceded their food sovereignty and sovereignty in general by allowing the exploration of their territories by this technology.
>
> (interview with Carrasco, 2012)

Acknowledging his inspiration from critical Latin American thinkers like Eduardo Galeano and Aníbal Quijano, Carrasco contextualized the fight against biotechnology as part of a continental fight for decolonization. The peasant movements, although framing the problems in a much more globalized way, also placed hopes in a continental fight in Latin America, seeking 'to interpret the political process at the continental level ... to be part of a space, a wider organization' (interview with SERCUPO, 2013).

The perception of the continental and global scope of the struggle against the agro-biotechnological model sharply contrasts with the nationally framed disputes over the policy for GMOs in Brazil, which will be told in the next chapters. However, as narrated in these pages, the recent years have shown an intensification of social mobilization in Argentina to dispute the national decisions that constitute the political economy of agrifood in the country as well as its neoliberal regulations on seeds, on chemicals and on the land issue. It has learned from other struggles, as with its neighbour Brazil, and started to dispute the national political decisions that maintain the model. These evolve in reaction to new needs of the pro-GMO coalition, opening new opportunities for social movements to reveal blind spots in their project of bio-hegemony and to shape an alternative food regime in Argentina that would better fit this theoretical concept, that is to say, that would produce food rather than commodities.

Notes

1 Svampa (2008) summarizes the meaning of those elections: they not only showed that the social mobilizations of 2002 were not translated into an effective challenge to political representation, but they also signalled the prospects of an infinite *peronismo*, as all other traditional parties had collapsed during the years of crisis.
2 At the beginning, isolated as they were, such groups espoused a reaction typical of other movements of neighbours who mobilize against the contamination of their

area with the installation of an industrial plant, a base for mobile phones transmission, an airport or a nuclear plant. Since they are concerned with the solution of their immediate problem, without elaborating further on the causes of the problem, such reactions are called NIMBY.

3 The neighbourhood, located 8 km from the city centre with circa 6,000 inhabitants, is over 60 years old and had never had basic public services such as water, sewage, electricity and paved roads.

4 The recovery of a protest repertoire that clearly marked the protest cycle of 2001, the *cacerolazos*, is to be interpreted with caution. Nevertheless, their spontaneity and lack of a defined ideology or claim cannot be considered as an articulated support to the rural actors but rather as a protest against the government (Svampa 2008, 230–231).

5 Although the president had discretionary powers to raise export taxes, the solution to the conflict was found by bringing the issue to the parliament. Resolution 125 was defeated in parliament.

6 'Sandra Ely Juárez died in front of a bulldozer … She intended to prevent them to raze the forests from the peasants. Cristian Ferreyra, … member of MOCASE-VC was killed on November 16, 2011 … Miguel Galván Lule-Vilela, also from MOCASE-VC, was killed by an employee of a soy entrepreneur' (Aranda 2013, 19).

7 Its founder, Enrique Viale, would later run as candidate of The Left Way (*La Via Izquierda*) for the national congress. His campaign relied on an environmental agenda but he did not achieve the minimum number of votes in the primary to compete in national elections.

8 In the final conclusion, it states that there are no sufficient and exhaustive studies on the long-term effects of glyphosate on health and environment under the conditions of use in Argentina, which it recommends should be done. Considering the existing data, it states that there is epidemiological evidence of the negative effects of pesticides on workers' health, such as skin irritations, cancer, abortions and malformations of their children, but there is no correlation between exposure and incidence of health effects, and that 'it is difficult to establish a relationship of cause-effect, due to interactions with environmental agents (generally mixture of substances), and genetic factors' (Comisión Nacional de Investigación sobre Agroquímicos and CONICET 2009, 129).

9 The invitation to Berlin was also a means of influencing the European policy because Germany is the rapporteur state in the EU for glyphosate. A restriction on glyphosate in Europe, in turn, could influence the international standards on the pesticide discussed at *Codex Alimentarius*.

10 For instance, in 2010, CONICET censored Carrasco's lecture at the book fair (Carrasco 2010).

11 This is how the proponents of the model attempt to frame the problem; this diagnosis would lead to the solution of educating the user. Indeed, this was the proposal of an organized reaction of INTA, AAPRESID, CASAFE, and other technical chambers, and agronomy universities (Todoagro.com.ar 2013).

12 Maybe it is more appropriate to interpret the Argentinean mobilizations not as a 'yet-to-be environmentalism' but as what they are. This is the proposal of some authors to understand these movements as socio-environmentalism (Bottaro and Sola Álvarez 2012; Svampa 2008, 2012).

Chapter 4
The unexpectedly contentious Brazilians
(1996–2002)

This chapter tells the story of how a controversy over GMOs was created in Brazil. Although the context of neoliberal adjustments in Brazil was highly favourable to the adoption of GM crops, and despite the efforts from a coalition of actors promoting the new technology, social mobilization against GM crops found a fertile ground among civil society in the 1990s due to the existence of organized movements among family farmers, peasants and agroecology activists. Another three reasons gave impulses to the dispute over agrobiotechnology. First, there was a divergence of opinions within the scientific community, which had a long tradition of engagement with the environmental movement. Second, this was a moment of consolidation of environmentalism in Brazil due to the economic boom and professionalization that surrounded the preparations for the Earth Summit in 1992 that took place in Rio de Janeiro, that is, civil society organizations were getting more professional in terms of applying for funding, managing public projects, hiring and paying qualified people. Third, civil society was keen on exploring the new legal opportunities for social mobilization and new institutional channels, which were opened due to democratization, for civil society representation at the executive level. These components provided the opportunity for scientists, Greenpeace Brazil and the Brazilian Institute for Consumer Protection (IDEC) to position agrobiotechnology as a precautionary issue and to generate public attention. They began a national controversy over GM crops. The watershed moment was the judicial action against the approval of GM soy that culminated in a legal moratorium. However, there were also precursors that politicized the technology in the south of Brazil, where GM soy was first cropped (illegally), thereby becoming the hotspot for the disputes between a pro-GMO and an anti-GMO coalition, spearheaded by the peasant movement. In 1999, the national campaign GM-Free Brazil was officially launched. All this mobilization on the side of civil society was met with an organized reaction on the part of biotech firms, state authorities and farmers who wanted the technology in Brazilian fields. This chapter narrates the main events that constructed a situation of controversy over GM crops in Brazil, which are depicted in Figure 4.1 below.

Neoliberal structural adjustments and a dual agrarian structure

After having successfully controlled hyperinflation with a programme for monetary stabilization in 1993, Minister of Finance Fernando Henrique Cardoso was elected president. The Cardoso government, from the Brazilian Social Democracy Party (PSDB), ruled Brazil from 1995 to 2002. This period was characterized by several reforms that reflected the Washington Consensus: macroeconomic stabilization, state

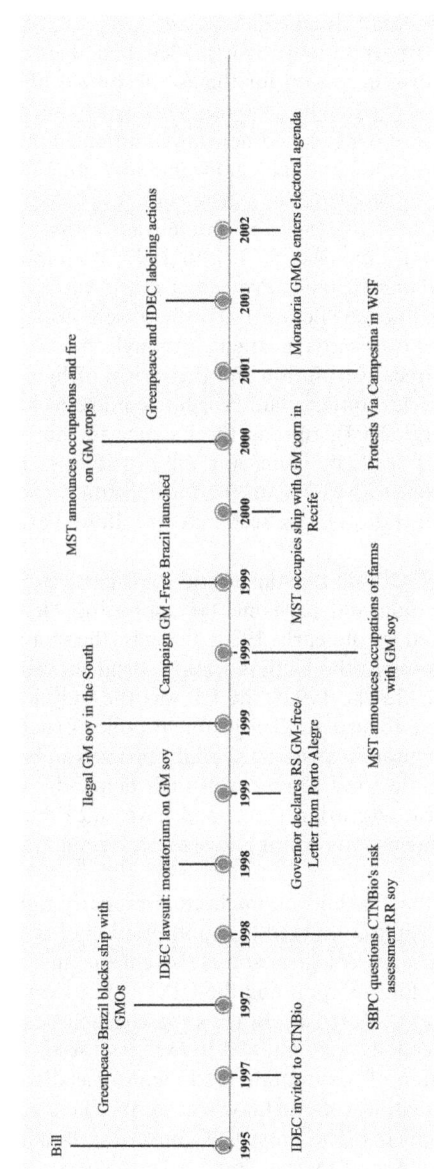

Figure 4.1   Timeline of events in social mobilization against GM crops in Brazil (1995–2002)
Source: Own creation.

reform, privatization and trade liberalization. The political and economic context was highly promising for the diffusion of genetically modified organisms. While farmers found themselves in a financially precarious situation due to the reduction of government subsidies, government was also in need of improving its trade balance. GM technology promised more profits and competitiveness, which met the interests of farmers and the government (Pelaez and Schmidt 2000).

The favourable context paved the way for the legal changes to incorporate the rules of neoliberal globalism necessary for the establishment of a corporate food regime. After having signed the free trade agreements to be a member of the WTO in 1994, the Brazilian government passed new laws and amendments to incorporate the new obligations into the national legal framework. In 1995, Brazil issued a *Biosafety Law* (Act 10,711). In 1996, the parliament translated the provisions of the *Agreement on Trade-Related Aspects of Intellectual Property Rights* (TRIPS) into the *Industrial Property Law* (Act 9,279); in 1997, its implications for seeds were secured under the *Plant Varieties Protection Law* (Act 9,504) and a project for a new *Seed Bill* was launched. Before 1997, there were no legal provisions for intellectual property rights over seeds in Brazil, although the seed market had been well-regulated since the Green Revolution and the arrival of hybrid seeds. The new laws thus lifted legal obstacles so that plant breeding could develop in the direction described by Kloppenburg (2004): tipping the balance to the private sector and to the commodification of seeds by increasing the possibilities for the protection of intellectual property rights as well as to the formalization of the seed market.[1] A strong process of concentration in the seed market followed after such laws were approved (Reis 2012).

None of this happened without resistance; and some concessions were achieved. There was social mobilization and parliamentary opposition led by the Workers' Party (PT). It was founded in the early 1980s through the convergence of social actors such as trade unionism, the Catholic Left, social movements (urban and agrarian) and intellectuals. In the 1990s, the PT was the only political party that played an oppositional role to structural adjustment policies such as privatization, state reform and macroeconomic austerity. Family farmers organized around the PT's Catholic Left elected deputies to represent their demands. These were organized in the Núcleo Agrário (Agrarian Core) and have since then been an ally in parliament for agrarian movements fighting for land reform and public policies for family farming.

During the voting of the new bills on intellectual property rights and plant varieties, the main point of dispute concerned the appropriation of seeds. The organized resistance of agrarian and agroecological social movements and agronomists influenced the Brazilian choice for incorporating the UPOV 1978 Convention instead of the UPOV 1991 Convention. Moreover, Brazilian law established a special protection for small farmers since they were allowed to exchange seeds among themselves (Reis 2012). The constitution of a corporate food regime was disputed by civil society. The agrarian transformation became known as agribusiness and was met with a reaction of agrarian movements. The project of converting Brazilian agriculture to an export-oriented corporate food regime would have to coexist with an alternative model based on very different ideological underpinnings. The peasants, once deemed to disappear with the advance of capitalism, instead developed a project for their own social reproduction.

Dualism in agriculture: agribusiness and family farming

Although there were actors and movements critical to the process that became known as 'agrarian modernization', promoted by the Green Revolution in the 1960s and 1970s, they could not find political allies. The structural transformations only increased the duality in the Brazilian countryside between big landholders undertaking large-scale production for exports and family farming (Delgado 2010). The former have historically been called *latifundiários* (*latifundium* are very large farms comprising extensive land use). The latter include heterogeneous actors such as rural workers, landless workers, smallholders, traditional communities with communal land, *meeiros* (contractual forms of land use) and *posseiros* (farmers who have been occupying a land for years and have the right to it but not the legal title). This dualism is reflected in different social categories, types of political representation, bodies in the state bureaucracy and public policies.

Big holders are known in their political activities as *ruralistas*. They are organized in a number of associations, among which the most prominent are the Brazilian Rural Society (SRB) and the National Confederation of Agriculture (CNA). They are represented in the parliament by the *bancada ruralista*, a cross-party coalition to defend their interests;[2] in the executive branch, their institutional spokesperson is the Ministry of Agriculture, Livestock and Supply (MAPA). Smallholders and farmers are represented by different organizations. Since the 1960s, the main form of representation has been rural unions such as the National Confederation of Workers in Agriculture (CONTAG), which brings together all subnational Agricultural Workers Federations (FETAGs), and the Federation of Family Farming Workers (FETRAF). Peasants fighting for access to land as well as rural workers demanding labour rights have had support from the Catholic Church, in particular after the creation of the Pastoral Land Commission (CPT) in 1975. These groups had no significant allies until the founding of the PT. In the state bureaucracy, their demands had been subsumed under the heading of land reform and land development; during the 1990s, the expression 'family farming' became established with the launching of the National Program for Strengthening Family Farming (PRONAF). State policies for these groups have been, in some periods, autonomous, in others, subordinated to the Ministry of Agriculture. Since the year 2000, the state policies for family farmers have been under the jurisdiction of the Ministry of Agrarian Development (MDA), created by the Cardoso government.

With democratization, a different context for social mobilization emerged, including the disputes around the model of agrarian development. Agrarian popular movements formed organizations and networks, engaged in new action repertoires and found allies among politicians and other social movements. They succeeded in including the issue in the new federal constitution of 1988, which states that land property is dependent on the fulfilment of a social function. This article was the beginning of a discursive transformation in the social representation of big landholders: they did not want to be associated with *latifúndio* anymore and therefore adopted a new label for their activity, agribusiness (*agronegócio*).

In the 1990s, agrarian popular movements would become prominent in the public sphere. Established in 1984, the MST became during the 1990s the largest grass-roots movement in the country (Carter 2010). In 1997, they organized a march to Brasilia with 100,000 people. The MST adopted the social movement form of political organization, and opted an action repertoire based on disruptive actions as well as land occupations, considered to be innovative in comparison to the then common forms of political representation and institutionalized action of the rural poor in workers' unions (Sigaud *et al.* 2008). The MST had support of the public and even some romanticized coverage in national mainstream media:

> Up to 1998 or so, our understanding was that the media had tolerated the movement[3] ... also because of the very strength of the MST, of contestation in that period. It was not agribusiness yet; it was backward landlordism. So the movement was ... the carrier of a modern message.
>
> (interview with MST, 2013)

The movement faced a much more favourable context when its main target was *latifundium*, which had no public legitimacy. But a new, more powerful enemy was in the making: agribusiness. Agribusiness became visible from 1998 to 1999, especially after the foreign exchange crisis in Brazil (Delgado 2013). The neoliberal policies led to macroeconomic stabilization at the high cost of leading to a crisis in the balance of payments as well as devaluation of currency in 1999. The production of commodities for export, which would generate foreign currency to cover the deficit in the balance of payments, became crucial for the government. Landholders started to use the term *agronegócio* to denote capital-intensive large-scale production, characterized by the high incorporation of technology, advancing the agrarian modernization model installed by the Green Revolution. They coined the term as part of a strategy to distinguish themselves from *latifundiários*, and from low-scale and low-capital production – thus highlighting the high capital investments and the image of businessmen. These transformations did not escape the attention of the agrarian movements, and their organic intellectuals (Sauer 2008, 2010), who spoke of a new stage of the 'conservative modernization' of Brazilian agriculture.

The Brazilian peasant movements already had very clear demands concerning public policy, which were not met by the government:

> [T]he Cardoso government ... created PRONAF ... as opposed to a policy for peasant farming that the movements demanded ... precisely to strengthen family farming, to create a sector within peasant farming itself that would be the *agro-negocinho* [the small-agribusiness].
>
> (interview with MST, 2013)

They aspired for the realization of another type of agriculture instead of following small-scale agribusiness. In this sense, they distinguish between family farming and peasant farming, reiterating that it is not the scale of production but the mode of production, that is to say, its relation to land and environment, which can be considered a special cultural identity. While maintaining their mobilization to create a food regime that would incorporate their proposals, many events took place in the urban settings by other social movements, scientific communities, state bodies and courts.

Precursors: scientists, environmentalists and consumers' movements (1990–1998)

The first debates over biotechnology in Brazil started in the academic community. In 1995, the First National Conference on Science and Technology voted for the regulation of biotechnology based on the precautionary principle (Cesarino 2006, 78). There was a close dialogue between scientists and the government, with the former being consulted to help with the formulation of the first *Biosafety Law*, issued by the parliament in 1995. However, the president vetoed some central articles, including the creation of the National Technical Commission on Biosafety (CTNBio). At the heart of the dispute was the consultative versus deliberative power of the commission: as the National Congress voted for the former option, leaving decision-making attributions to already existing competent authorities in the areas of health and environment, influent scientists lobbied the government to veto this article (interview with Nodari, 2013).

Meanwhile, the Brazilian Society for the Advancement of Science (SBPC) and the Brazilian Society of Genetics (SBG) started to organize the first debates on biotechnology, beginning in 1997. Scientists who followed the topic recall that the SBPC during the four presidential mandates encompassing the offices of Sérgio Ferreira (1995–1999) and Glaci Zancan (1999–2003) were marked by a critical position demanding precaution in regards to GMOs. These representatives of the scientific community often took positions in public debates by publishing their views (Folha de São Paulo 2000a). 'There were also individual statements, some for, some against, but the majority demanded more studies and a critical assessment, but were not necessarily against it' (interview with Nodari, 2013). Some of these scientists would later be members of the CTNBio, marking their participation by demanding further studies. Also, many scientists formed the first wave of environmental activists in Brazil (Hochstetler and Keck 2007).

Another group of experts involved in the issue had a very different profile from those scientists conducting laboratory research, being very concerned with regulations for research and commercialization. These were agronomists interested in the practical consequences that biotechnology would have for agrarian development and for farmers on the ground. Their technical base was guided by stated political and ideological definitions (Pelaez and Schmidt 2000). Since the 1970s, they have been critically studying the effects of the Green Revolution in the country as well as proposing agroecology. During this time, they have formed various associations and NGOs to support the reproduction of family farmers as a social category and the prevention of farmers' dependency on agrarian inputs from international firms.[4] Pelaez and Schmidt (2000) argue that resistance to GMOs in Brazil can be better understood as a new struggle from this movement. This movement created a network of organizations that promoted farmers' autonomy and control over seed production. Their engagement explains the fundamental role played by organizations of family farmers moved by the idea of autonomy in the GMO dispute. On another front, they put pressure on law projects and policies in the areas of pesticides, seeds and intellectual property rights in the 1990s, thereby achieving some success, as mentioned earlier. Their critique of agrobiotechnology and their proposals regarding an alternative agrarian development reached a wider audience – not yet concerned with what happened in the agrarian fields – when GM crops was taken up as an issue by the environmental and consumer movements.

Environmentalism and Greenpeace Brazil

Greenpeace was the very first to take a direct action against GM crops in Brazil, not least because it was one of the first actors to have access to information about the technology arriving in the country. Its decision to act was influenced by its organizational capacity and evaluation that they could undertake a successful campaign against GMOs. In 1992, the Brazilian office was founded following the Rio conference, a moment in which environmentalism was at its peak in the country. The environmental movement established networks with many movement sectors and was experimenting with different forms of participation in politics. Greenpeace Brazil was grounded in the national reality, in which environmentalism could not avoid taking positions on issues of social justice. This resulted in the discourse of socio-environmentalism in the 1990s (Hochstetler and Keck 2007). From the late 1980s, the deforestation in the Amazon had centralized the movement's agenda. Greenpeace Brazil evaluated that GM soy provided a good entry for the organization to change the stage of their actions and connect distant forests and seas to the urban landscape (interview with Greenpeace Brazil, 2012).

Because of its transnational scale of operation, Greenpeace became aware of the arrival in Brazil of a shipment with GM soy. This was requested by the Brazilian Association of Vegetable Oil Industries (ABIOVE) and approved by CTNBio in 1997 – which was operating despite its uncertain legal status. The first direct action ensued: Greenpeace activists blocked the shipment with transgenic soy. Given the high prominence of the organization also in foreign and international news agencies, together with its novelty in Brazil and its difference from local environmental organizations, Greenpeace caught the national media's attention. It was reported on *Fantástico*, a TV show with a large audience on Sunday evenings, and from then on the topic started to get noticed (interview with Greenpeace Brazil, 2012). The organization relied on its repertoire of spectacular events to raise media attention.

In addition to media-designed protest events, Greenpeace engaged in legal mobilization, motivating the public prosecutor to undertake public civil action. Freitas (2011) notes that legal mobilization is not a common tool used by Greenpeace, but the reason for this exception was Greenpeace's evaluation that, at this stage, it would have had immediate effects that probably would not have been achieved by protest actions alone due to the emergency of the situation. It was the quickest way to stop GM crops from arriving in the country, until they started a public campaign. An injunction was granted demanding ABIOVE respect the Code of Consumer Protection and label the products produced with GM soy (Salazar and Grou 2010). The demand for compliance with existing laws or for the establishment of appropriate legal framework coincided with the zeitgeist of democratic consolidation, in which the law and its exercise became a part of contentious politics.

Legal opportunities and participatory democracy: the consumer rights movement

During the 1980s, the Public Prosecutor's Office started a reform through which its criminal prosecution profile gave space to civil suits oriented towards law enforcement in the protection of public interests and collective rights. This tendency was strengthened by the *Law of Public Interest Suits* (Act 7437) in 1985. This is a legal instrument that ascribes responsibility for damage to the environment, consumer goods as well as rights of artistic, aesthetic and historical value. It gives legal standing to public interest organizations, which can either engage directly in judicial battles or motivate the Public Prosecutor's Office to act, being a more strategic choice due to the power and resources of the latter.

The IDEC was founded in São Paulo in 1987 with a very strong law profile but also technical expertise in specific areas. The IDEC aimed to go beyond individual judicial actions and started proposing class actions (*ação coletiva*), using the new legal opportunities available with the democratization of the country. At the core of the organization's action repertoire is legal mobilization, but it is used in combination with activities to raise awareness such as courses and campaigns of information and orientation of consumers as well as participation in the legislative process and in executive bodies. The decision to take up action against GMOs did not follow the demands from its associates as it was a very abstract issue; rather, it was the result of the institutional designs of participatory democracy in Brazil combined with the recognition of the IDEC's reputation. Because of its expertise in food issues and its strong representative role in official bodies, the organization was invited in 1995 to represent civil society at CTNBio.[5]

The agenda of the commission was the authorization of research and commercialization of RR soy (interview with former official IDEC, 2013). The IDEC became aware of the issue through this invitation. It did not participate for long, however, as its members noted that instead of creating rules for health and environmental risk assessments the commission's aim was to liberalize GMOs, following the USA and Argentina:

> When we went to check the experimental fields where CTNBio had authorized research activities before the commercial approval, they were experiments in areas of 100, 115 hectares, what characterizes seed multiplication. They had no dimension of experiment fields, but aimed at the liberalization that was about to happen.
> (interview with former official IDEC, 2013)

Because of this, the IDEC issued a formal statement asking to leave the commission. The aim was to be independent in order to contest the entire process: the illegal operation of CTNBio, the lack of norms for risk assessment and the approval of GMOs.

The unexpected moratorium and the rift in the state (1998)

On 24 September 1998, CTNBio issued a favourable technical opinion on RR soy, disregarding the objections raised during the past meetings by the SPBC and the IDEC (Leite 1998a, 1998c). The SBPC had made official its position by sending the commission a questionnaire on technical issues including allergenicity and the toxicity of glyphosate. On the following day, with the imminent governmental approval of GM soy for commercial production, the IDEC introduced an injunction, supported by the public prosecution, to suspend the approval of RR soy: 'We demanded an injunction to prevent liberation until norms were created to evaluate risks to health and environment and a labelling rule to guarantee consumer rights. These were the three issues and the ground was the absence of norms' (interview with former official IDEC, 2013).

The injunction was granted on the same day, derailing the planned approval of GM soy. It functioned as a legal moratorium for five years and was in force until 2005, when the *Biosafety Law* suspended it.[6] It was a legal moratorium but as events unfolded it proved not to be a de facto one. Notwithstanding, this did not change the

IDEC's assessment that this lawsuit was the first big victory of social mobilization against GMOs:

> Media interest started with the novelty of the issue, and with the judicial battle ...
> A small organization, a consumer organization, moves a lawsuit against the federal
> government, which soon received the support from Monsanto ... and it managed to
> prevent the liberalization.
>
>   (interview with former official IDEC, 2013)

Thus, the judicial action provided good news coverage, with a big dramatization of a conflict between unequal parties, a consumer organization versus a global giant – a script such as 'David vs Goliath' (Fernandes 2010). It was covered in the mainstream newspaper *Estado de São Paulo*, and reached a wider public, reinforcing the public debate initiated by Greenpeace actions. Many activists remember this moment as a golden era on the issue (interviews with AS-PTA, 2012; former official IDEC, 2013; Nodari, 2013).

The judicial action affected the script from the biotechnology firms in Brazil and its coalition with government sectors and scientists from CTNBio (Salazar and Grou 2010). For this reason, the moratorium marks the official start of the contentious politics over GM crops in Brazil, made effective by the judiciary branch. The ruling did not prohibit GM soy but worked as a suspension of its approval, based on the absence of studies on its environmental impact. The implications are spelled out by an activist: 'Either Monsanto conducted the environmental impact assessments or it invested in the fight in the judiciary. It preferred to invest in the fight rather than to conduct the studies' (interview with AS-PTA, 2012).

Thus, Monsanto avoided regulatory and public scrutiny over its scientific data by focusing on the judicial action. It entered into the lawsuit siding with the Brazilian state in defence of CTNBio. Greenpeace sided with the IDEC together with the Brazilian Institute of Environment and Renewable Resources (IBAMA), the regulatory agency from the Ministry of Environment. The executive power requested the general attorney to issue its opinion, which must prevail in case of disagreement in the government.[7] The representative from the IDEC interpreted that as a sign of the pro-GMO policy that the government was planning to establish: 'This shows how the government was involved in the issue, supporting, wanting the liberalization, politics utterly disregarding the technical level' (interview with former official IDEC, 2013). It also provides evidence of the weakness of the environmental legal framework, always under threat of being neglected and thus in need of permanent vigilance from environmental movements (Hochstetler and Keck 2007).

The judicial battle opened up internal rifts among different ministries that would continue for many years to come: on the one side, the ministries of environment, health, agrarian development and justice (where the department of consumer rights is located); and on the other side, the ministries of agriculture, science and technology, industry and trade, and foreign relations. While networks of activists for environmental, health, consumer and peasant rights were part of or were represented in state bureaucracy (Hochstetler and Keck 2007, 19), networks for a bio-hegemonic project also infiltrated the state. Faced with the unexpected halt by a judicial action, the Ministry of Agriculture did not accept this overthrow of its plans. It had been negotiating with Monsanto the product launch in the country for three years (Leite 1998b). Despite the impossibility of making RR soy legally available for the season of 1999/2000, the Ministry of Agriculture urged Monsanto to register it as a plant variety anyway. At the same time, the Brazilian Corporation of Agriculture Research (Embrapa), which is subordinated to the Ministry of Agriculture, kept multiplying RR soy seeds, following a cooperation agreement with Monsanto (Rippardo and Murakawa 2000). Brazilian farmers on the border to Argentina started to voice their demands to try the new technology after hearing from their neighbours about the benefits. They started to smuggle the GM seed, also in the *bolsas blancas*, in other words, without paying the price for intellectual property rights. Already in October 1998, GM soy illegally planted was confiscated and the producers were fined by the police. A polarized conflict started in Rio Grande do Sul over GM soy.

On the border to Argentina: farmers-turned-smugglers

By the year 1999, Argentina had converted its soy production to agrobiotechnology, crossing the threshold of 70 per cent of cultivated soy. Argentina is much closer to Rio Grande do Sul than to Brasilia, where official decisions on GMOs take place. The border also is more than a metaphor for the margins of state authority: it is easily crossed and escapes state vigilance (Das and Poole 2004). The year 1999 started with the first mandate of a state governor from the PT and an increased number of seats in the legislative. With close links to family farming, peasant movements and agroecology, the governor Olívio Dutra declared the intention to make Rio Grande do Sul a GM-free zone, with stricter terms than the national legislation (Fernandes 2010; Pelaez and Schmidt 2000). His statements were widely covered in the media, both regional and national (Gerschmann 1999), unleashing a heated controversy.

Enjoying this favourable political context, a coalition including organizations from environmental and peasant movements, agroecology cooperatives, trade unions, lawyer associations and consumer agencies groups, as well as pastoral groups issued the first public statement from civil society in Brazil demanding a public debate on GMOs on 20 August 1999: The Letter from the Rio Grande do Sul. They demanded the suspension of approvals of production and imports of GMOs and the funding of public research on risks. They called other Brazilian and world regions to join them in building a world free of GMOs ('Carta do Rio Grande do Sul' 1999).

In the meantime, soy producers were trying to find their own way to use GM soy. The seeds smuggled from Argentina became famous as *soja Maradona*, in reference to the renowned Argentinean soccer player and his outstanding performance. They were already monocropping soy and had been experiencing pest management problems; thus, GM crops appeared as a modern solution. Not only did they manage to get seeds by illegal means but they also disputed the legitimacy of their actions: the Agriculture Federation from Rio Grande do Sul (Farsul), representative of medium-sized landholders, became a strong advocate for the use of GM soy among its members and defended it publicly. In 2000/2001, producers caught with illegal soy had their crops incinerated. However, producers kept breaking the law and GM soy was estimated to cover two-thirds of Rio Grande do Sul soy crops in the harvest of 2002/2003 (Embrapa Soja 2003). Based on an ethnographic research conducted in 2000, Menasche (2003) notes that farmers differentiated between their own production and consumption. While they were attracted to the promised advantages of GM soy, such as cost reduction and labour saving in pest management, and they were ready to apply pesticides if they were to increase yields, they refrained from using pesticides on the crops of their own consumption. She identifies 'an inverted work ethic' as the motivation for farmers to adopt the new technology. Their only concern was losing control over seeds and their practices of exchange; nevertheless, this fear was not enough to deter them from the pressure from other farmers to adopt the new technology.

The polarization over GMOs in southern Brazil raised scholarly interest (Lima 2007; Menasche 2003, 2005; Silveira 2004). Lima (2007) identified two groups, with contrasting conceptions of agrarian development. Techno-optimists defended a techno-productive model as established by the Green Revolution; eco-socials contested it and defended agroecology. The former was dominant (with values like modernity, progress, scientific authority, the power of markets); the other challenged the status quo, trying to show that another model is possible. This was more than a dispute over GMOs, it was an ideological and power dispute.

Contentious peasants' movements

The growing adoption of GM soy by family farmers increased contestation over the technology. Already in 1999, the MST announced a plan of occupation of farms that planted GM seeds in Rio Grande do Sul for the next two years (Samora 1999). Activists from Rio Grande do Sul highlighted the GM crops issue during the wave of protests organized by the MST known as Rise of the Countryside (*Levante do Campo*). In Recife, the MST engaged in another type of direct action not part of its protest repertoire: together with activists from the Central Workers' Union (CUT), it occupied a ship in the harbour with the intention of burning its cargo, GM corn imported from Argentina (Guibu 2000). The plan against GM crops was officially nationalized two months later, in the IV National Congress of the MST and received wide media coverage, which highlighted statements related to disruptive actions such 'to occupy farms' and 'to set on fire' GM crops (Silva 2000).[8] Farm occupations soon would become a reality, especially in the southern states of Rio Grande do Sul and Santa Catarina, where illegal GM crops were spreading. Among the concerns of the MST was how the technology would impact rural employment due to the very same argument used by its proponents, that is to say, labour saving.

The state of Rio Grande do Sul, however, would remain a main stage of the resistance to the new technology. In January 2001, Porto Alegre hosted the first World Social Forum (WSF), in which the PT and the state government played a fundamental role by officially joining the conveners of the event and financing half of its costs. During the event, there were important multi-scalar actions against GM crops. Via Campesina announced a petition to the United Nations asking for the recognition of non-GM seeds and creole seeds as human heritage (Folha de São Paulo 2001a). José Bové, an activist from the French organization Confédéracion Paysanne (Peasants' Confederation), participated in a joint protest with the MST. He attracted national and international attention from a wide range of media since he had already become a famous media figure due to his use of disruptive action during the protests against globalization in Seattle in 1999. About 1,000 activists from the MST participated in a protest against GM seeds. They uprooted GM soy plants from experimental fields run by Monsanto, where they burned seeds and the US flag. The leadership announced they would burn all fields with GM crops in Brazil and demanded that the national government stop Monsanto from installing an industrial plant in the state of Bahia or they would occupy it and prevent its operation (Editorial 2001; Traumann 2001). The federal police, following superior orders, notified the French activist to leave the country and he used habeas corpus in order not to be arrested. This caused a popular media story, exposing how the federal government aimed at intimidating the activist but without making it public – in vain – in order not to be considered to be siding with Monsanto (Barros e Silva 2001). This was a symbolic act that brought together the transnational peasant movement Via Campesina with the movement for global justice.

Agrobiotechnology entered the international agenda of the Via Campesina in association with seeds. There have been long debates on the issue since the late 1990s. A leader from the Movement of Small Farmers (MPA), an organization that has been active in the GMOs issue, explained that it is characteristic of Via Campesina to quickly take up emerging issues in order to debate them and define a position. The meetings provide for the exchange of experiences, their interpretation and the definition of strategies (interview with MPA, 2012). Via Campesina incorporated the issue of GM seeds under the overarching master frame of food sovereignty, proving its developed articulational scope as a problem-solving scheme that is flexible enough to include and interpret new events in the world:

> In the food sovereignty debate, …, when the GMO issue comes, the first concern is, well, here is a gateway for losing control of our seeds. Because it is not that they are going to make us buy their seeds, but is that our seeds will be contaminated.
>
> (interview with MPA, 2012)

Food sovereignty was constructed in response to the discourse of food security while prioritizing farmers' rights over seeds, considered to be threatened by the adoption of GM crops. Within this master frame, the peasant movements elaborated on the effects of GM seeds in terms of environment, health, politics and economy; they bridged all issues, extending the original core of the right to land. This framing process taking place transnationally under the auspices of Via Campesina was well fitted to the structural transformations taking place in the Brazilian agriculture, helping movements to make sense of it. They could critically interpret the material and institutional changes in the power of landholders and agrarian corporations that accompanied the discourse on food security. In fact, agrobiotechnology was incorporated as one more issue in this long cycle of protest from peasant movements that had started with democratization and got harder with neoliberal structural adjustment policy. Via Campesina Brazil had a strong grass-roots mobilization capacity as it comprises a network of ten organizations in Brazil related to land and agricultural issues.[9]

The beginning of the GM-Free Brazil campaign

The imminent approval of GM soy and the success of the injunction suspending it provided 'the big boom, the big alert signal' (interview with AS-PTA, 2012) that motivated other organizations to join efforts and launch the campaign Por um Brasil Livre de Transgênicos (GM-Free Brazil). The original coalition was formed by nine organizations from various areas: Consumer rights (IDEC), environment (Greenpeace, Centro Ecológico), agroecology and family farming (AS-PTA, Esplar), public policy monitoring (Institute for Socioeconomic Studies, Inesc), communication (Center for Creation of Popular Image, Cecip), community development (Federation of Organs for Social and Educational Assistance, FASE) and development aid (Actionaid).[10] These organizations were formally linked by means of a funding project from Novib, a cooperation agency from the Netherlands, which already financed the NGO Advisory and Services for Projects in Alternative Agriculture (AS-PTA) in various projects. The initial funding was used to start the communication activities and to organize the meetings. For the first years the campaign had an NGO profile.[11] It used legal mobilization, communication and mobilization as fronts of action (interview with AS-PTA, 2012).

The legal front was for a long time coordinated by the IDEC. By then, legal mobilization was already central for the campaign. The information front was assumed by the AS-PTA. It was founded in 1983 to promote sustainable rural development, focusing on family farming and agroecology. The AS-PTA organizes a weekly newsletter (*Boletim*) consisting of an editorial with its point of view regarding the situation of GMOs in Brazil and in the world, with news clippings with the link to its source, sometimes followed by a paragraph commenting on it, and an example of agroecological alternatives. It also publishes regular updates in English and sends them to allies in other countries. Besides the newsletter, the organization produces materials for campaign mobilization and monitoring reports of the activities from CTNBio, all of them are posted on the website. The sources are mostly news media and official reports, but they also include press releases from social movements. The *Boletim* became a reference for critical information on the issue not only for activists but also to sympathizers in governmental agencies and media groups. The communication front thus tailored its messages to different audiences: 'Information was produced for farmers, journalists, consumers, students, technicians and congress members in the form of texts, leaflets, newspapers, videos, audio reports and opinion pieces for newspapers' (Fernandes 2010, 2).

Finally, the mobilization front was in charge of organizing direct action: protests, seminars, distribution of flyers, posters and broadsheets. In May 2000, the campaign, under the head of FASE, organized a seminar with the Heinrich Böll Foundation, from the German Green Party, in Rio de Janeiro. It was called Transgenic Food: International Alliance for a Moratorium. There were representatives from the German Green Party, and their allies in the Brazilian legislative, from the Secretary of Agriculture of Rio Grande do Sul, MST, CONTAG, and other activists, and scientists. Another example of action was the Tribunals of the People on the Impacts of GMOs, organized by the GM-Free Brazil campaign together with peasants and small farmers' associations. It had various editions, in 2001 in Fortaleza, Belem, in 2003 in Porto Alegre. The goal was to bring information and stimulate public debate on the issue. The method followed the model of a court, with a judge, a prosecutor, a lawyer for defence, and a jury, composed by small farmers and urban consumers appointed by urban and rural trade unions (interview with Terra de Direitos, 2013). The campaign worked until 2003 with this composition. Functioning as an informal network, organizations were free to keep their own preferred strategies, as will be described next.

Targeting capital: an urban alliance between environmentalist and consumer movements

Greenpeace worked strategically on its framing choices, as GMOs made the organization go well beyond its primary interests: environmentalism. Transgenics became an issue where the environmental discourse could be combined with health and consumer rights' frames. This has been one of the campaigns through which Greenpeace achieved higher resonance in the urban sector. Indeed, this frame bridging allowed Greenpeace to make GMOs an issue with salience in the concrete daily lives of urban dwellers as food consumers by relating it to their phenomenological world. Establishing a relation between biotechnology and their daily food gave some experiential commensurability to the topic, a powerful factor in face of the abstractness and complexity of the technology:

> This issue is too complex for us to restrict it to environment; it is cross-cutting ... Of course that our main discourse was environment, that there were various environmental risks that the technology that was said to be environmental friendly was in the long-term a failed technology.
>
> (interview with Greenpeace Brazil, 2012)

This was not only a discursive strategy. It received empirical credibility, providing energy and strength to the framing activities, as Greenpeace launched an action repertoire of what they called corporative work, which scholars defined as 'targeting capital' (Schurman and Munro 2009). The organization tested food products for the presence of GM material; in positive cases, they sent letters to the responsible industry stating that consumers did not want that, and also staged protests in front of industry's headquarters. They created and distributed lists of industrialized food that contained transgenic ingredients. These lists caught the attention of mainstream media many times (Folha de São Paulo 2000c, 2000d, 2001f, 2002a, 2002b). Greenpeace targeted soy farmers by protesting in GM fields, which also reached the media (Folha de São Paulo 2001d). Activists undertook awareness-raising activities in supermarkets, where they collected signatures for petitions (interview with Greenpeace Brazil, 2012).

Meanwhile, the IDEC had developed a campaign for labelling GMOs, including a draft bill open for public comments (Scheinberg 2000), and various strategies to raise awareness for their associates: a handbook in 1999, magazine articles, and information events on the International Food Day in 1999 and 2000. Among the associates, there was 'a very good reception, as the topic was published in almost every issue of the magazine, and later came the tests, which had a huge resonance' (interview with former official IDEC, 2013). The IDEC also helped keep the issue on the media agenda, due to two action repertoires: legal mobilization (Folha de São Paulo 2000b) and corporative work, contracting laboratories to detect the presence of GM material in supermarket products (Folha de São Paulo 2001b; Leite 2000).[12]

The IDEC thus acted as an alert system, early on diagnosing GMOs as a problem for the consumer. At the National Forum of Consumer Protection Organizations (FNEDC), it created a booklet and disseminated it across Brazil. Transnationally, in Consumers International, a federation of national consumer rights organizations, the Brazilian campaign was a successful case:

> [T]o introduce this debate ... using several different strategies, ranging from infor-
> mational work, consumer mobilization, acting on official bodies and the judiciary,
> it was a lot ... especially being successful, preventing the release. ... Brazil ... played
> undoubtedly a very important role in the international movement.
>
> (interview with former official IDEC, 2013)

Greenpeace and the IDEC hired the Brazilian Institute of Public Opinion and Statistics (Ibope) to conduct surveys on public opinion in 2001, 2002 and 2003 to gauge public acceptance of GMOs (Freitas 2011). The conclusion that the more informed they were, the more people rejected GMOs provided further support for their actions. The convergence of the adopted framings prepared the ground for a promising and new alliance in Brazil between consumers and environmental movements, with an innovative action repertoire that politicized the market sphere, including consumers, producers and firms (Freitas 2011).

Building up the legal front and the auspices for a wide coalition between urban and agrarian flanks

In 2002, the newly founded human rights organization Terra de Direitos joined the legal front of GM-Free Brazil, which has remained strategic for the campaign as there are many judicial actions being processed in the judiciary. The year 2002 was also when the *Jornadas de Agroecologia* from Via Campesina Brazil were organized for the first time, in Curitiba, Paraná, with the topic 'Land free from GMOs and from pesticides'. Terra de Direitos, also based in Curitiba, participated and took some messages home:

> Genetic contamination stood as a concrete issue in the lives of those groups that had
> seeds, and we understood then that we should work with this problem like two sides
> of a coin. On one hand, to defend farmers' right to have their seeds; on the other,
> begin to work with GMOs, to monitor the releases.
>
> (interview with Terra de Direitos, 2013)

The topic of GM crops arose already under the frame of farmers' rights over seeds and the master frame of food sovereignty, and entered the agenda of Terra de Direitos in a bottom-up process, out of tangible problems faced by farmers and peasants. Terra de Direitos combined such grass-roots contacts with their insertion into transnational advocacy networks for human rights. They not only strengthened the front of legal mobilization in GM-Free Brazil but also brought additional repertoires of action while targeting foreign and transnational audiences.

The joining of organizations working in the countryside such as Esplar, AS-PTA and Terra de Direitos incorporated themes such as agroecology and family farming into GM-Free Brazil. The multiple sectors involved in the campaign provided the basis for a dialogue between the urban campaign targeting the market and focusing on consumer rights and agrarian frames related to agroecology and farmers' rights in the process of constructing a critical interpretation of GM crops. An often mentioned outcome of the first years of the campaign was the alliance with the PT, which exercised a consistent oppositional role in national politics. During the electoral period, GM-Free Brazil managed to include in the programmatic statements of presidential candidate Luís Inácio Lula da Silva (Lula) a moratorium on GMOs until there were conclusive studies on the risks.

The pro-GMO coalition

The social mobilization against GMOs provoked reactions from opponents and authorities. Although resistance found allies in subnational governments, the federal government was highly engaged in promoting the technology. Actors interested in the approval of GM crops in Brazil strengthened their coordination activities, also forming a coalition, characterized, though, by different strategies. These aimed at increasing their material power over plant research, the seed market and agrarian production; their institutional power by strengthening the coordination among all interested parties in the government and in the private sector; and their discursive power by investing in public relations and media, and by establishing censorship of dissidence. In short, there was a project of bio-hegemony.

Activists recall that Monsanto, in collaboration with state institutions, pursued a strategy of controlling the seed market in Rio Grande do Sul. In fact, the GM seeds not only came from Argentina but were also produced in Brazil, which was a result of a contract between Monsanto and Embrapa signed in 1998 (interview with AS-PTA, 2012). Embrapa would contribute with its bank of seeds, having accumulated varieties suited to the different climates and soils of Brazil, and Monsanto would provide its patented genetic modification, the insertion of the gene RR. Embrapa spoke openly about the agreement and defended biotechnology at any opportunity (Murakawa 1999). Moreover, it issued an internal memorandum prohibiting its employees to publicize their contrary views on biotechnology and adding a document with the official position of the body defending GM crops (Rippardo and Murakawa 2000).

Embrapa did not invest in seed multiplication of non-GM soy adapted to the Rio Grande do Sul, a problem still faced by farmers in that state. This added to the reasons why more farmers adopted GM seeds, legal uncertainties notwithstanding: they had no alternative offer of conventional seeds (interview with AS-PTA, 2012). Other farmers, however, were purposively taking the legal risk because they bet on the governmental approval. They had at least certainty that the national authorities would not prevent them from doing so. Relying on the lack of enforcement, they increased their bets every season. GM soy spread to the states of Paraná and Santa Catarina, and reached the states of Mato Grosso do Sul and Mato Grosso, in the centre-east of the country. By 2002, 20 per cent of the soy fields in Brazil were estimated to be GM; from these, 70 per cent were located in Rio Grande do Sul (Cardoso 2002).

In terms of media relations, Embrapa published a paid article occupying an entire page in a prominent space of the economy section of the most read mainstream newspaper in Brazil on 4 November 1999: 'The importance of biotechnology for Brazil' (Sociedade Rural Brasileira and Embrapa 1999). It stated the benefits of GMOs to the country. It was co-signed by the SRB, the Brazilian Organization of Cooperatives (OBC), the Brazilian Organization of Plant Breeders (BRASPOV) and the Brazilian Association of Seeds and Seedlings (ABRASEM).[13] The president of the Brazilian Rural Society also published an opinion article advocating GMOs (Hafers 2000). At the end of 1999, a journalist who would later become the science editor of *Folha de São Paulo* published in a news article that the pro-GMO coalition had sent a video and paid advertisements to most major newspapers on the advantages of the new technology (Leite 1999).

In January 2000, the *Cartagena Protocol* was signed. Officially, Brazil was part of the like-minded group formed of developing countries from the G-77 and China. They envisaged a strong protocol for the protection of biodiversity. However, the

Ministry of Agriculture informally allied with the position of the Miami Group. Thus the Brazilian national position was an object of internal disputes and would continue to be in the future negotiations of the Cartagena Protocol. Among the main points of dispute were rules of labelling and segregation. Brazil supported an intermediary position, which neither conformed to the interests of the G-77 for a strict system nor those from the Miami Group for no labelling at all. In the end, Brazil became a party to the protocol that recognized that GMOs were different from other non-modified organisms and that regulation should be based on the precautionary principle and on risk assessment, which foresaw mechanisms of public participation, transparency and prior informed consent in the international trade with GMOs. This implied the obligation of incorporating such provisions into national law. At the same time, Brazil served as an insider ally of the Miami Group – whose member countries were not parties to the protocol – in their constant efforts to deter the strengthening of this biosafety regime. At that time, Brazil was internationally observed as having the decision on whose side the balance would tip in the GMOs issue (Folha de São Paulo 2000e).

In July 2000, the federal government issued a note signed by six ministries, led by the Ministry of Agriculture (Folha de São Paulo 2000f), in defence of biotechnology as a priority for the progress of the country. In the opposition, deputies from the PT, led by the environmentalist politician Marina Silva, threatened to install a Parliamentary Inquiry Commission on the governmental pro-GMO position (Rippardo 2000). In 2001, the executive power issued decrees in hopes of accommodating both sides of the conflict. On the one side, it ordered labelling of all food products containing more than 4 per cent of genetically modified material in their composition (Decree 3,871). The parliament transformed it into a law, which entered into force the same year. This threshold, however, did not satisfy the IDEC (Folha de São Paulo 2001h). Nevertheless, it served as a basis for the social movements' campaigns, which then focused on monitoring the application of such law with tests and publicity.

On the other side, it altered the *Biosafety Law* (Provisional Measure 2,191) by providing legal existence to the CTNBio, subordinated to the Ministry of Science and Technology and validating its past administrative acts. It stated that the decisions of the CTNBio were binding to the other governmental bodies, exacerbating a long political dispute and a legal controversy inside the executive power concerning the division of labour between the new body created for biosafety issues and the traditional decision realm of the Ministry of Environment. In 2002, upon request from the Ministry of Science and Technology, the federal attorney issued an opinion stating that, in case of shared competences between the CTNBio and other federal institutions, the CTNBio had the final say over the existence of risks related to GMOs (n. AGU/MP-02/02, annex to GM-032, Process n. 00001.006775/2001–78). The Ministry of Agriculture understood that with those measures all legal obstacles had been removed to the definitive approval of GM soy and announced it would issue a decree authorizing it (Gerschmann 2001). However, the IDEC and other organizations from GM-Free Brazil kept guard within the judiciary to maintain its ruling until environmental assessments were conducted, as well as organized protests (Folha de São Paulo 2001g).

Their corporative work on food industry also set off an organized reaction. The Brazilian Association of Food Industry (ABIA) denied all information published by the IDEC and Greenpeace, claiming that as there was no GM soy in Brazil it was not possible that GM ingredients were found in processed foods, as well as that there was no appropriate methodology to test it. Some industries, like Nestlé, ordered control tests in other laboratories and contested the results published by the NGOs (Folha de São Paulo 2001c). In 2002, the pro-GMO coalition established an NGO, the Council of Information on Biotechnology (CIB). The main transnational biotech firms are among its associates (Bayer, Basf, Dow, DuPont, Monsanto and Syngenta), as well as ABIA, ABRASEM, and research institutes and universities. The NGO hired a public relations agency in order to influence public opinion on GMOs (Lyra 2005).

The year 2002 ended with the election of a new national government, which would inherit the imbroglio of the controversy over GMOs. During his electoral campaign, Lula travelled to the south where he criticized the then governmentally stated intentions of authorizing GM soy (Folha de São Paulo 2001e). Soon after the election results, the USA announced all its support for the food security programme that was part of the electoral agenda from the PT: Fome Zero (Null Hunger). Monsanto used it as an opening to establish contacts with the future government. This was the beginning of the conversion of José Graziano, the coordinator of Fome Zero at the time, to the project of bio-hegemony, and with him, of many politicians within the next government, as will be told in the next chapter.[14]

Notes

1 Reis (2012) explains the differences between formal and local seed systems. Registered, certified and patterned seeds compose the formal system; local systems are characterized by farmers' traditional uses of saving, selecting and exchanging seeds. Instead of a linear evolution form the latter to the former, Reis affirms that they have coexisted and that they intersect: farmers may complement their own seeds with commercial ones, and seed developers use farmers' seeds as a basis for their product development. The use of their own seeds has remained significant in Brazil, despite the increasing legal measures (such as conditions for receiving state funding) to make farmers buy commercial seeds.
2 Despite their informality, they constitute one of the most influential and organized interest groups in parliament, backed by a strong agrarian lobby. They not only nominate the presidencies of the legislative commission for agrarian policy but also dispute the leadership of any other commission that might affect their interests, such as the environmental commission.
3 On 17 April 1996, the state police killed 21 activists from the MST in the state of Pará, in what became known as the 'Massacre of Eldorado dos Carajás'. Since then, the MST organizes a series of protests every April in what they have called 'Red April'. 'The *Eldorado dos Carajás*, with all the international solidarity, ended up putting the movement in a condition that the media could not make us invisible' (interview with MST, 2013).
4 Pelaez and Schmidt (2000) explain that this is the origin of the NGO Advisory and Services for Projects in Alternative Agriculture (AS-PTA), which would later play a central role in the organization of the GM-Free Brazil campaign. They formed the Projects in Alternative Agriculture (PTA), associated with the Federation of Organs for Social and Educational Assistance (FASE). Later it became autonomous as Rede PTA, and then it transformed into the NGO.

5 Food has been an important focus of the IDEC since its very beginning. Its first judicial action, taken in 1988, demanded that the Ministry of Agriculture prohibit the use of hormones in livestock farming. After that, important food issues were milk quality, misleading nutritional labelling, dietetic products and fraud with frozen products. Their work on GMOs follows this tradition.

6 The opposite happened in the European Union: a de facto moratorium against GM crops was in place from 1998 to 2003, although it was not de jure. These mismatches between reality and legality are not a monopoly of Brazilian contentious politics over GM crops.

7 In 2002, it decided that only the CTNBio had the prerogative to express the official position on biosafety; IBAMA was forced to withdraw from the action.

8 Note that the newspaper speaks of 'invasion', whereas the movement calls their actions 'occupations', based on the constitutional right to land reform on lands that do not fulfil their social function.

9 These are: MST, National Movement of People Affected by Dams (MAB), Movement of Peasant Women (MMC), Movement of Small Farmers (MPA), CPT, Rural Youth Pastoral (PJR), Movement of Quilombolas (CONAQ), Movement of Brazilian Fishermen and Fisherwomen (MPP) and Indigenous Missionary Council (CIMI). There is a degree of organicity by means of an annual plenary meeting in which priorities are defined, and a national coordination that works along the year. But each movement also retains its autonomy. When there is consensus regarding a common agenda, the movements closer to the issue represent Via Campesina.

10 During this period, other organizations also participated in the campaign, among which the Embrapa Workers Union (Fernandes 2010), showing that there was dissidence and critical views on the technology inside this governmental agency subordinated to the Ministry of Agriculture.

11 This was a moment of professionalization of civil society in Brazil, with the formation of NGOs that mobilized legal, administrative and scientific expertise, and of institutionalization, as NGOs increased their cooperation agreements with the state to execute projects and also to represent civil society.

12 These actions took place during 2000 and 2001 and had to be abandoned due to lack of funding.

13 Established in 1972 by various local and regional associations of breeders, today it has among its members all multinational seed companies. Its membership overlaps with BRASPOV.

14 In 2011, he was elected as president of the Food and Agriculture Organization.

Chapter 5
The politics of *fait accompli* (2003–2013)

After their victories in creating a controversy over GMOs in Brazil and achieving a legal moratorium, activists were faced with a tough reaction from the actors' network defending GMOs. From an illegal crop, GM soy was legalized by the National Congress and became a dominant crop. It did not take long for Brazil to be side by side with the pioneers in the use of technology. This chapter tells the story of the political struggles behind such an abrupt transformation. It is divided in two parts. The first covers the period from 2003 to 2005 and comprises the first years of the new government, when social movements tried to influence the decisions on how to solve the issue of illegal crops and on the new biosafety law. The main battles were fought in the executive and legislative arenas, which were very polarized, and culminated in a pro-biotechnology decision. This victory entailed a redefinition of social movements' targets and strategies: it was no more about influencing society and debate regarding the best legal framework for GM crops, because this was a lost battle; now, the first and foremost priority was to influence the next locus of decision-making, namely the commission of experts to find some room for precaution. This will be narrated in the second part, which deals with the period from 2006 to 2013. From 2008 onwards, as the pro-GMO policy converted the country into a GM-full Brazil, movements campaigned to address its consequences, such as contamination and pesticide use. During these years in which the biotech food regime consolidated its material, institutional and discursive powers, social movements never ceased to contest the dominance of GMOs in Brazilian fields.

Shaping reality and legality (2003–2005)

The year 2003 and the subsequent two years were crucial for the GM-Free Brazil campaign because of a constellation of five factors. Until then, the campaign was comprised mainly of NGOs with a professional profile and less social movement bases and grass-roots networks. This changed with the official inclusion of agrarian movements, which, in turn, benefited from the alliance with scientists engaged in the campaign. A second factor was the expectations over the new president, who, while promoting in his electoral programme a moratorium on GMOs, entered into office with the pressing problem of the illegal cultivation of soy in Rio Grande do Sul. Third, as the problem divided the government along the existing faults in the executive branch, it moved the battle to the parliament. In 2005, the *bancada ruralista* managed to legalize GM soy, settling the GM crops issue as a politics of *fait accompli* and making Brazil an official producer of GM crops. At the same time, and here lies a fourth element of this period, a process of criminalization of peasant movements was under way, which, together

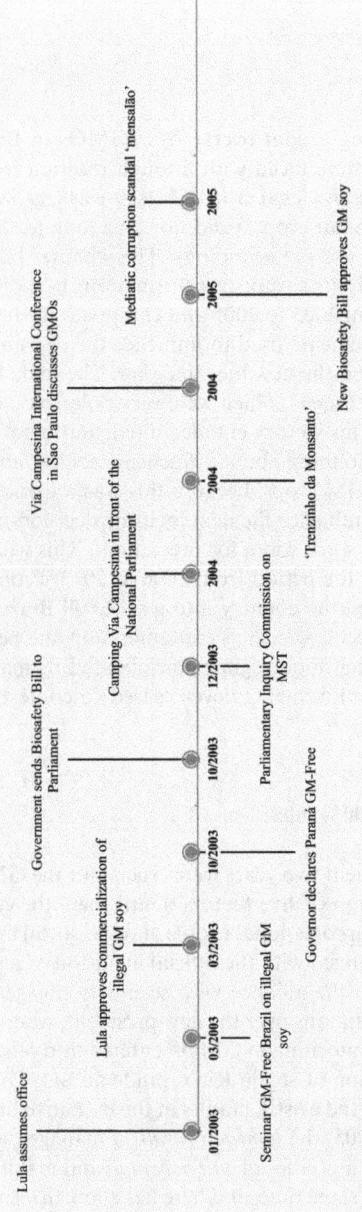

Figure 5.1 Timeline of events in social mobilization against GM crops in Brazil (2003–2005)
Source: Own creation.

with other processes, contributed to their relative demobilization at the time. A final factor characterizing this phase is the consolidation of agribusiness. Figure 5.1 summarizes the events in these years of intense controversies over the Brazilian GM policy.

*Enlarging the campaign: peasants fighting agribusiness
and scientists as allies*

For various social movements who helped to elect Lula, the election of the Workers' Party meant that their political allies assumed office in 2003. At first, social movements were empowered with a sense of 'it is our government, now we can pursue our policies'. Indeed, many activists were invited to assume office in state bodies or widen their access to public funding for their organizations in the provision of public services, such as the educational activities of the Landless Workers' Movement. The president received the MST in the very first year, showing openness for debating their agenda. Lula assured them 'that in his government there would not be any soy, no liberalized GMOs' (interview with MST, 2013).

Lula assumed office with the issue of illegal GM soy crops pending in the political agenda. In the second month in office, the government created an interministerial commission on GMOs, in which the divisions inside the government became evident: the ministries of agriculture, trade and development, science and technology on one side, and the ministries of environment, health and agrarian development on the other (Cesarino 2006, 37). The new government did not manage to settle the prevailing disputes among ministries, by failing to give a new political orientation. In addition to these conflicts, activists anticipated the contradictions that could arise due to the programmatic commitment made during the electoral campaign of fighting hunger and the destruction of soy.

This moment is considered a turning point in the GM-Free Brazil campaign, when activists organized a seminar called 'The Threats of GMOs: Proposals from Civil Society' that took place in Brasilia in March 2003. It had two main goals: to widen the campaign to include farmers and to propose a solution to Lula regarding the *soja Maradona* in Rio Grande do Sul. The mobilization of family farmers was seen as the means to guarantee grass-roots resistance to GM crops, or else lose the battle, as the case of Rio Grande do Sul had presaged. In order to counter the reports on the advantages of GM seeds by Argentinean farmers, American and Canadian farmers were invited to testify regarding their negative experiences with the same product, warranting experiential commensurability of the identified threats (Fernandes 2010, 4).

This seminar marked the entry of peasant movements into the GM-Free Brazil campaign. Via Campesina Brazil had a strong grass-roots mobilization capacity and a critical position on the issue since the late 1990s. In 2003, it launched a global campaign on Seeds: Heritage of Rural Peoples in the Service of Humanity, and in 2004, during the IV Conference of Via Campesina International, which took place in São Paulo, representatives from countries where GM crops had been adopted recounted their experiences (interviews with MPA, 2012; MST, 2013). Similar to Greenpeace Brazil, Via Campesina Brazil decided to engage in the GM-Free Brazil campaign as a result of a combination of transnational incentives and national/local realities.

At first, the GM crops issue was of regional, not national concern for peasant movements, especially because they were confident that Lula would not allow GM seeds in Brazil (interview with MST, 2013). On the subnational front of resistance in the south, there was an escalation of opposition to GM crops in Paraná, the neigh-bouring state to Rio Grande do Sul. In 2003, the anti-GMO movements gained an ally in the head of that state. Elected in 2003, the state governor, Roberto Requião (Brazilian Democratic Movement Party, PMDB), was openly against GM crops. He prohibited the import and export of GM seeds through the state's harbour, Porto de Paranaguá. The governor passed a law declaring Paraná a GM-free zone. These actions provoked the states of Mato Grosso and Mato Grosso do Sul, which had expe-rienced an immense boom in soy production and were interested in the new technol-ogy. They became open supporters of GM crops and entered into a subnational battle with the state of Paraná by bringing a lawsuit that declared the actions prohibiting the transport and cropping of GMOs to be unconstitutional (Tortato 2003). The Supreme Court judged it to be unconstitutional according to the argument that the federal gov-ernment is the sole decision-maker concerning GMOs. Paraná would challenge the national policy for years to come.

The entry of Via Campesina brought political strength to GM-Free Brazil nation-ally due to its transnational reach; Via Campesina made the Brazilian campaign globally known. This also benefited peasant struggles: 'With the debate on GMOs, through Via Campesina, we managed to give visibility to the outside, in Europe, of how perverse the model of agrarian production in Brazil is' (interview with MPA, 2012). In addition, its participation matched the programme of NGOs working with agroecology such as AS-PTA. While most of the organizations forming the anti-GMO movement, both nationally and globally, mobilized in an issue-oriented way, peasant movements constructed a political identity that allowed GM seeds to be framed in an antagonistic fashion. Building on a daily basis another model of agriculture that is oriented towards food sovereignty, they challenge agribusiness and its expressions, such as GM seeds, as well as organic production. Their political identity as peasants is related to a mode of production that entails not only producing without GM seeds and pesticides but also producing in an agroecological way.

Last but not least, peasant mobilization against GM crops paved the way for attracting allies among scientists:

> With GMOs, we managed to attract a group of researchers who did not participate in the debate on the struggle for land. They do not take part; they are not concerned. But when it comes to research, to genetic contamination ... all this more scientific debate ... it brings in other partners, who are also concerned and who also see an ally in the peasantry, in the agrarian social movements.
>
> (interview with MPA, 2012)

GMOs catalysed this alliance, according to the testimony from Rubens Nodari, a pro-fessor of genetics at the Federal University of Santa Catarina. Nodari realized that he was part of a group of scientists who defended the precautionary principle and who shared the same concerns with others in society, such as small farmers and agroecolo-gist NGOs. He started to collaborate with them by giving courses, helping to detect seed contamination and establishing 'participatory research' projects in areas such as genetic improvement and agroecology.

Peasant movements recall how scientists helped them understand the technical aspects of GMOs (interview with MPA, 2012). For their part, scientists like Nodari clearly differentiated between knowledge asymmetries related to the need of a specific scientific training to debate some technical issues, and other types of knowledge, including farmers' knowledge over seeds and environment, as well as the more general democratic competence of civil society to discuss the many other issues related to GM crops. Such recognition denies the transference of legitimacy from science to politics:

> [W]hen we discuss the technology, it actually requires a basic, relatively large, knowledge of genetics and molecular biology. But then, that is the second point, this issue is multi-disciplinary ... there are the socio-economic, cultural issues, there is biodiversity loss ... and civil society does have the means to participate because she knows much about these issues.
>
> (interview with Nodari, 2013)

The politics of fait accompli

The seminar organized by GM-Free Brazil in 2003 had a second goal: to formulate a civil society proposal with legitimacy ensuing from the participation of small farmers, and for the government to solve the issue of the illegal GM soy in Rio Grande do Sul. The solution they arrived at was to isolate the harvest and export it, and thus start the next season with conventional seeds (interview with AS-PTA, 2012). They were aware of the political, economic and discursive constraints faced by the federal government to punish illegality by destroying the crops. The government disregarded the proposal and edited a provisional measure that approved the national commercialization of those crops (MP 113). This measure 'radically changed the course of history in terms of GMOs in Brazil' (Fernandes 2010, 3). Activists recall it as the first big defeat of Marina da Silva, by then a world-known environmentalist,[1] who assumed office as minister of environment, creating high expectations among social movements (interview with Greenpeace Brazil, 2012). In this way, the government signalled its weakness to the pro-GMO coalition. The weakness to enforce legislation is not meant as a function of state strength; rather, it would be best understood as a demonstration of a political will that prevailed, which however cannot be expressed unambiguously as such. The pro-GMO coalition did not wait to take these signs as opportunities to advance their intentions to transform Brazil into a producer of GM crops through illegal means; however, they were preparing a strategy to win the legal battle.

When the next sowing season arrived, in September 2003, farmers started pressuring the government to liberalize GM seeds, stating that they would go bankrupt otherwise. The activists from GM-Free Brazil identified that there had been a strategy of depleting the stock of conventional seeds by mixing them with the grains to be exported because soybeans exported from Rio Grande do Sul were rejected by import authorities in China due to high levels of fungicide.[2] Farmers argued that there were no conventional seeds in the market, creating a situation in which they were 'forced' to use GM seeds.[3] Close to the following harvest, farmers from Rio Grande do Sul, represented by the FETAG, again pressured the government to liberalize commercialization. A second provisional measure was issued, and a third one followed later in the year. With these provisional measures, the executive power attempted to bring reality into legality: illegal GM soy cultivation was authorized *a posteriori*, creating what social movements called 'politics of *fait accompli*' (interview with AS-PTA, 2012).[4] At the same time, the acts from the executive power had an uncertain legal status since they ignored the judicial action in force that conditioned the authorization of GM soy on the realization of risks studies. Now the constitutional powers were in an open conflict. The executive remained divided. In the judiciary, another judge had conceded Monsanto a preliminary injunction that authorized RR soy, which was later reversed by the Regional Federal Court. Moreover, the Supreme Court was involved in three judicial actions by the Green Party (PV), the CONTAG and the Public Prosecutor's Office, claiming the unconstitutionality of the provisional measures (Cesarino 2006).

Meanwhile, the legislative maintained its interests in the issue by organizing public hearings and proposing various bill projects (Taglialegna 2005). The agrarian lobby demanded from the government a permanent solution to the problem and the *bancada ruralista* threatened to vote on the provisional measure and convert it into a law, approving GM soy *tout court*. Faced with these pressures, the government sent the draft law project PL 2,401/03 to the parliament (interview with AS-PTA, 2012). This is how the pro-GMO coalition managed to bring the issue into their sphere of influence, by relying on the power of the *bancada ruralista* in the parliament. The next battle moved to the National Congress.

The battle moves to the National Congress: the debates over
the Biosafety Bill

The years 2004 and 2005 were marked by a debate about the new *Biosafety Bill*. Both camps in the fight organized themselves to influence the process. During this period, Via Campesina, led by the MST, organized a protest by camping in front of the National Congress for about 40 days. For GM-Free Brazil, this marked the actual entry of peasant movements into the campaign. They mobilized their bases and invited congressmen, scientists and other organizations to their camp. The negotiations of the *Biosafety Bill* attracted the support of the Fórum da Terra, which includes religious organizations such as the CPT and the National Confederation of Brazilian Bishops (CNBB), trade unions (CONTAG and CUT), scientists and the public health research institute Fundação Oswaldo Cruz, public agencies on consumer protection (Procons) and the Public Prosecutor's Office. GM-Free Brazil won the support of legislators that were closer to the movements and provided more critical positions. They established the Parliamentary Coalition in Defence of Biosafety and the Precautionary Principle, headed by João Alfredo (PT later PSOL),[5] and composed of 80 representatives from the parties PT, PV, PMDB, PSDB and Brazilian Socialist Party (PSB), including the Núcleo Agrário from the PT, environmentalists and legislators motivated on various grounds. This coalition counted on the support of the Ministry of Environment (Cesarino 2006, 83; Taglialegna 2005, 43, 44).

The *bancada ruralista*[6] found new allies in the biotech lobby and the scientific community. The traditional interest groups fostering agribusiness included the CNA, OBC, SRB, and organizations from the seed industry, ABRASEM and BRASPOV. Organizations from soy producers' states joined the coalition: FARSUL (Rio Grande do Sul), FAEP (Paraná) and FaMASUL (Mato Grosso do Sul). Dissidence among the workers' union was marked by the entry of FETAG from Rio Grande do Sul into the pro-GMO coalition. From the scientific community, the most active organization was the National Biosafety Association (ANBio), as well as the Brazilian Science Academy (ABC) and individual experts. Institutes that conducted research on GMOs, such as Embrapa, Central Cooperative of Agrarian Research (COODETEC), Agronomic Institute from Paraná (IAPAR) and Fundação Mato

Grosso, also joined. In addition, the coalition received support from the ministries of agriculture and science and technology. Biotechnological firms (Monsanto, Novartis, AgrEvo and Pioneer) used other associations as informal channels (Cesarino 2006, 83; Taglialegna 2005, 43, 44).

Monsanto was also involved and promoted what became known as the '*trenzinho da Monsanto*', in reference to a train ride in an amusement park. A strategy used by Monsanto in other countries (Schnurr 2013), it was not directly promoted by the firm; rather, it was organized with the cooperation of the United States Agency for International Development (USAID), thus with the active involvement of the US embassy, and paid for by ABRASEM, which Monsanto is a member of (Valente 2003). They are called 'seeing is believing' tours, in which politicians, journalists, civil servants and also NGOs are invited to visit the firms' headquarters in the USA and South Africa.

In this situation, activists identify a rupture that would be one of the two crucial difficulties activists faced in the negotiation of the bill: the split in the position of the PT regarding GM crops (interview with AS-PTA, 2012). What had been a programmatic position in the PT, a precautionary public policy towards GM crops in line with social movements' demands, only clearly displayed the positioning of the Núcleo Agrário. These deputies maintained loyal to social movements, especially in Deputy Adão Preto (Rio Grande do Sul). Other PT deputies started to polarize the party.[7] Activists attribute it to electoral funding by agribusiness. According to Cesarino (2006), the 'lack of cohesion' in the position of the PT as a party and of the governmental coalition as a whole in the negotiation and voting of this bill is explained, in part, by the subject matter itself as well as by the difficulties in the relationship between the government and its congressional basis. There was a regionalization of the debate, with deputies from states of high or potential soy production participating more actively in the debate. More than party position, which was very unclear, the main determinant for the National Congress members' voting choice was their relation to interest groups: siding with agribusiness on the one hand, or with environmentalists on the other.

The second major obstacle faced by GM-Free Brazil was the enlargement of the scope of the law to include stem cell research (interview with AS-PTA, 2012). For activists, this was a successful strategy from the pro-GMO coalition to derail the debate about GM crops, placing 'smoke and mirrors' in the parliamentary arena. It functioned as a red herring fallacy, when one topic not relevant to the issue on the agenda is introduced under the guise of being part of it and the new topic leads to the abandonment of the original issue. What was to be a *Biosafety Bill*, clearly created to resolve the controversy over GM crops became known as 'the stem cell bill'.[8] Mass media responded to the strategy as expected, covering mostly aspects of medical research and ethics. The pro-biotechnology coalition invited people with disabilities wearing T-shirts with the slogan 'stem cells, cells of life' (interview with AS-PTA, 2012). The combination of two very different issues brought difficulties for deputies who had contrasting positions regarding the technologies. Both sides of the controversy were aware that this strategy of maintaining the stem cell topic in the scope of the *Biosafety Bill* would help the approval of both technologies, one bringing votes to the other. In hindsight, activists confirmed their assessment that it was a false promise, judged from the relative absence of deliberation on stem cell research in the Brazilian regulatory agenda.

The higher resonance of the stem cells debate did not preclude the many polemics on GMOs that marked the negotiation of the bill. These can be grouped in two types. The first concerned the role of the National Technical Commission on Biosafety in decision-making (consultative versus deliberative); consequently, its actual power in relation to regulatory agencies in the areas of health and environment, including its competence to decide if an environmental assessment was necessary. The GM-Free Brazil campaign defended a consultative role, justified in terms of accountability: only state institutions responsible for health and environmental risks should be entitled to decision-making since they can be held accountable for consequences. The counter-reaction of the pro-GMO coalition was to frame the proposals from GM-Free Brazil as anti-science, again by relying on a technocratic ideology in which science provides legitimacy to politics and competitiveness for the economy. This was not merely an opposition of technocracy versus democracy, since the GM-Free Brazil campaign defended that the decision be taken by the existing technical governmental bodies competent in risk assessment, the Brazilian Health Surveillance Agency (ANVISA) and the IBAMA. However, the pro-GMO coalition disagreed with the incorporation of the technical expertise from these governmental bodies because of their institutional location, which is outside the sphere of influence from *ruralistas* as well as their reputation for defending collective rights. The better they do their job, the more they threaten dominant interests for which risk assessments pose obstacles to economic activities. As noted by Hochstetler and Keck (2007), the environmental institutional and legal framework is always under threat and legal attributions might be usurped. Indeed, the key issue for the pro-GMO coalition was to create a commission under their influence and with the superpower to override IBAMA and ANVISA, since the *ruralistas* are 'terrified' of them (interview with AS-PTA, 2012). Their position prevailed. The bill also created a superordinate body, the National Biosafety Council (CNBS). The council is composed of 11 state ministers and subject to the office of the president of the republic, designated as the higher advisor agency for formulating and implementing the *National Biosafety Policy*. It was meant to be a high political body, responsible for issues of national interest as well as social and economic impact, whereas the CTNBio was designed to be a technical, multidisciplinary collegiate of experts. Activists found it an interesting model (interview with AS-PTA, 2012).

The second node of controversy involved the rules to guide the functioning of the commission. GM-Free Brazil, through its alliance with the Parliamentary Coalition for Biosafety and the Precautionary Principle, managed to include some articles that would pave the way for the new phase of their campaign: the monitoring of the CTNBio. Nevertheless, the overall result of the negotiations in the high chamber was the text of the *Biosafety Bill*, which incorporated again most demands from the pro-GMO coalition.

The negotiations of the *Biosafety Bill* involved many stages, passing in special commissions in both chambers, and going back and forth between the chambers. In this process, the text suffered modifications, pending between the extremes of the actors for and against GMOs. The ambiguities in the official governmental position came to the surface in each of these moments. There was never a clear orientation for deputies (Cesarino 2006, 47, 65, 106). For Lisboa[9] (2007, 2009), the PT used various strategies to avoid making an open statement on a change of its prior positions to avoid confronting the Ministry of Environment and the social movements that supported its government. The party had no common position to instruct their Congress members nor to ask for support of the allied parties; thus, in the name of governability, the government yielded to the pressures of the pro-GMO, which had a majority. According to Cesarino (2006, 68), parties like the PV, PSOL and PCdoB (*Partido Comunista do Brasil*, Brazilian Communist Party) were also divided on how to vote, tending to favour stem cell research but having a more precautionary approach towards GMOs. Christian deputies had the opposite position. There was also a convergence of positions between government and opposition. In sum, the legislative process revealed the inconsistencies in the government regarding the policy for GMOs and this is why Cesarino (2006) argues that the National Congress was the arena in which the negotiations shaped the *Biosafety Law*.[10]

The president issued the bill with some small changes, disregarding the appeals by the ministries of environment and health. The main concession to their demands was to pass a decree on the voting rules of the CTNBio, demanding a qualified majority, which displeased very much the pro-GMO coalition. This would at least provide some room for dispute in the expert commission. The public prosecutor intervened with an injunction of unconstitutionality (Adin no 3526), with the support of the PV and IDEC, contesting the powers of the CTNBio to decide on commercial authorization and the need for environmental licence.

The *Biosafety Law* had various implications. First, the law cleared the legal status of GM soy, putting an end to the effects of the lawsuit from 1998 (the moratorium). GM soy was never evaluated in terms of health and environmental risks and the National Congress legalized the politics of *fait accompli*. The law thus marked a reversal for GM-Free Brazil in terms of policy outcome. It was a watershed event as it changed the context of action for the campaign:

> [U]ntil the law, there was the legal issue, the question of information, debate, to win more forces in society for the discussion. And after the law was issued, that there was the regulating law, now the context is different, now we have to turn the focus to those who will make decisions.
>
> (interview with AS-PTA, 2012)

However, the overall evaluation of the first years of the campaign was positive as it reached public debate, had media attention – which has waned since then (interview with former official IDEC, 2013) – and articulated a network of different movements and organizations working on a common front. The latter is a very rare accomplishment: 'I think that in that period we have brought together a lot of organizations in a very short time … fighting for a single issue. And this is not very easy, knowing the history of our left' (interview with MPA, 2012). However, it was not only the left that was united; the conservative forces increased their power and mobilization as well, in particular with the commodity boom, and coordinated an offensive against peasant movements.

Criminalization and demobilization of peasants' movements

During these first years of Lula's term, the movements of Via Campesina faced numerous difficulties for mobilizing their grass-roots bases against GM crops. These can be traced to four main reasons: a process of criminalization of peasant movements by state institutions and the media; the political space occupied in the government by strong figures of agribusiness sending the message that there were few chances for peasant movements to dispute the agrarian policy; the difficulty in mobilizing their constituencies against a government identified as their representative and from whose policies they perceived real benefits; and, last but not least, the adoption of GM crops among their own farmers.

The process of criminalization of social movements related to the fight for land worsened at the beginning of Lula's office. During the electoral process, as Lula's chances of winning the presidency rose, the MST recalls a clear change in the media, which characterized their activists as rioters and looters. 'After the first two years of the Lula government, the MST had a high number of occupations, and the Brazilian elite was in fact concerned with land reform' (interview with MST, 2013). In 2003, the *bancada ruralista* established a parliamentary investigative committee to inquire into the activities of the MST (Lupion 2005). Another such initiative followed in 2006 when oppositional parties established a commission to investigate the transference of public resources to NGOs, with the goal of attacking the alliance between the PT government and social movements. These inquiry commissions brought the MST into the public debate with the purpose of spreading a negative image of the movement and, with it, criticize the governing party for its alliances. They were accused of stealing and diverting public money to practice 'vandalism' and 'invade' land and private property. They were accused of being 'a gang', a 'militant group taking advantage of the Lula government': 'This is precisely the period that we define as the beginning of the criminalization of social movements. It is a period of invisibility on the part of media, and of criminalization' (interview with MST, 2013). Surrounded by such difficulties and having to retreat to their main struggle – land reform – the MST reduced its actions against GM crops.

A second factor of demobilization was the closure of political space to conduct a change in the agrarian policy. The nomination of an intellectual from agribusiness, Roberto Rodrigues, to head the Ministry of Agriculture showed that the government was not going to change the model of agrarian development. To the contrary, all clues pointed to the consolidation of agribusiness under his government. The proximity to China in the foreign relations also was a sign of the export-led growth based on commodities (Sampaio Jr 2012), among which were soybeans. Connected with the increase in demand, a factor in conjunction contributed to this export-led growth strategy: a boom in the international prices in commodities, including soy. However, this was not merely an economic issue. Social movements have also attributed the consolidation of agribusiness to political factors, namely to the type of alliances needed to maintain the federal government. In 2005, there was a so-called 'problem of governability', emerging from the wave of attacks suffered by the PT due to the uncovering of a corruption scheme in the parliament that became known as *mensalão*. The criminalization of peasant movements was exacerbated in that year as the government faced its strongest attacks from the media and the opposition. There was a shift in the political alignments in the direction of a more conservative agenda. All promises to make effective land reform were broken (interview with MPA, 2012). The core of governmental policy being decided, social movements had to work at the margins, in the interstices left – that is to say, by increasing the dialogue with the Ministry of Agrarian Development – that did not challenge the core, being the *ruralistas* and the Ministry of Agriculture. The peasant movement organizations also professed a *mea culpa*. In hindsight, they believe they had misinterpreted the structure of political opportunities: believing it to be a government under dispute, their strategy was collaborative and patient in order to avoid weakening the government vis-à-vis all the conservative forces that were betting on its failure. They believed that some spaces, such as that of land reform, were guaranteed in the government. After the first years, they realized that it was a government of compromise (interview with MPA, 2012).

In those spaces that Lula's government prioritized, such as social policy, it did achieve important policy results. Here comes the third factor: demobilization due to social conformism. Despite the clarity in the diagnosis from the social movement leaders that the Lula government consolidated the agribusiness model, which represents their main enemy in the fight for food sovereignty, the organizations from Via Campesina faced great hurdles in mobilizing their bases against the national government due to the general perception that the government was good to them. As extreme poverty is concentrated in the rural population, cash transfer policies targeted at the very poor have benefited the countryside, as Lavinas (2012) recalls. According to her, what contributed most to the reduction of rural poverty was the policy of a maintained increase of the minimum wage.[11] However, when considering non-monetary indicators, the rural poor continue to lack access to the most basic services. Therefore, the policies targeted at bringing electricity to the rural areas had a huge impact:

> they never had electricity in their lives, and the government takes office and offers you the possibility of watching television at home, of having a refrigerator ... then how will you explain this to a comrade that this government ... has a policy against you? It is a debate that was very difficult for us. And this ensured ... a kind of social conformism.
>
> (interview with MPA, 2012)

The fourth factor explaining the difficulties of peasant movements in joining the efforts from GM-Free Brazil is the adoption of GM crops by their grass-roots, as already mentioned. This was a major weakness because peasants are the ones who produce crops daily. With the official approval of GM soy, public agrarian extension services started to promote the big novelty. 'The first argument was: You are going to produce more, and spend less, use less poison. When one touches the economic aspect, it catches on' (interview with MPA, 2012). After all, farmers are also economic agents and arguments on costs and productivity are also part of their value system. During these first years, they experienced immediate economic benefits, bringing credibility to the claims defending GM seeds.

Silencing public debate and building bio-hegemony

During these three years, the pro-GMO coalition advanced on various fronts. They won the legislative battle to approve GM soy; they succeeded in intimidating peasant movements; and they increased their sphere of power in the federal government by consolidating agribusiness among the state priorities. They also managed to influence mass media and to decrease the degree of controversy over GMOs.

The work from Castro (2006) provides evidence for the argument that the efficiency of the new technology does not explain its adoption but that the institutional arrangements involving economic, political and social aspects do. Based on media material that traces the social networks of actors engaged with legalizing GM soy,[12] she highlights three groups of actors responsible for its diffusion. First were the farmers who illegally planted GM soy because they believed it would reduce their costs. Second, the patent holder biotech firm, Monsanto, developed a series of strategies. In terms of material power, it reduced costs to gain clients, developed a royalty collection system, and made acquisitions to concentrate power in the national seed market, while internationally it sought supply control by influencing all major soy producers to turn to GM and thus restrict the power of consumer demand in import markets. In terms of institutional and discursive power, it made use of marketing strategies such as publicity, funding of NGOs, conferences, workshops, publications and informal meetings to form social networks in support of the technology. Finally, the government ignored judicial rulings and failed to punish illegal activities. Because it lacked a political orientation, the national government contributed to foment the controversy among subnational units, judicial battles and also the illegal conduct of farmers. Moreover, the government's behaviour tended to collaborate with Monsanto's lobby due to, among other reasons, the high investment capacity of the firm.

Lisboa (2007, 2009) names two reasons for the decrease of controversy over GMOs in Brazil: the drop in media coverage and the support of the PT. She recalls that the media covered the issue intensively from 1998 to 2004 when – coinciding with a public relations campaign from Monsanto in print media, television, cinema and outdoors – media groups reduced the space dedicated to the issue, gave less standing to social movements, and repeated the arguments of biotech industries without checking their claims.[13] Regarding the second factor, the support from the Lula government, Lisboa argues that despite the high expectations placed on his office concerning the advancing of an environmentalist agenda it often meant a regression regarding that of Cardoso. The government supported agribusiness for two reasons according to Lisboa. First of all, it was a strategy of political opportunism, of exchanging support in one area for advancement in other areas. While the *bancada ruralista* supported Cardoso's initiatives, Lula could never be sure of their support and had first to ensure their neutrality so as not to block his measures. He used 'political arguments' on the convenience of not opposing the Rio Grande do Sul farmers and important governors from the PMDB, taking the issue as being of minor importance and which could be given to the rural lobby in exchange for their support (or their abstention) on other issues considered to be priorities of the government. At the same time, and this is the second reason, the option for agribusiness coincided with the conception of development adopted by Lula's government. The highest rank in the party shared a developmentalist ideology, according to which environmental and social justice are obstacles to economic growth, identified as development, although this conception opposes the principles from its social bases.

Lisboa's analysis on the reasons for Lula's government approval of GMOs coincides with the more general theses from political scientists on Lula's government. Singer (2012) defends a thesis of 'weak reformism', according to which Lula focused on social policy and avoided other issues that could be confrontational with the status quo. Nobre (2013) defends the argument that since redemocratization of the country the conservative forces have vetoed any proposals that could jeopardize their interests; as a result, the government in office faced the challenge of searching for spaces that avoided such veto power. These tendencies became even clearer towards the end of 2006 with the presidential elections. Due to a major oppositional campaign against the PT, with the help of media groups that magnified evidence of the corruption scheme in 2005, there was a shift in the ruling coalition in the name of 'governability'. The PT allied with the PMDB, which nominated the vice-presidential candidate and important ministries, including the Ministry of Agriculture, since agrarian elites are the historical constituents of that party. This resulted in the formalization of *pemedebismo*[14] into national politics and also symbolized the change that the PT was undergoing since it had started its upward trajectory to federal power.[15] This signalled the official closure of the executive power to the demands of GM-Free Brazil. Even state agencies that had historical ties to the environmental and health movements, serving as allies of the campaign, were silenced by the political decision to favour agribusiness. It weakened its leftist factions that espoused more radical goals regarding social change. Some PT members, and defectors who formed a new left party (PSOL), continued to support the GM-Free Brazil campaign. Oppositional parties with higher electoral results offered no prospects for social movements since they represent agribusiness (DEM, the Party Democrats) and neoliberal agendas (PSDB). With the legislative and executive branches on the side of the rural lobby, only the judiciary power remained a possible ally in the state.

When uncertain futures become true (2006–2013)

The reversals in the policy and legal framework for GMOs, under a government that they helped to elect, made many social movements and organizations from GM-Free Brazil interpret the structure of political opportunities in a more critical way. Some organizations understood that the fight was lost. The campaign became smaller, led by the IDEC, AS-PTA and Terra de Direitos. This does not mean that the issue lost legitimacy as an issue to fight for. It was rather a matter of resources that made some organizations discontinue a consistent monitoring of the issue. But when it came to specific moments, other organizations participated in protest activities. The AS-PTA attributed this to the trust that was built among organizations in the campaign. To them, the fight had now really started, as 'all impacts that we are forecasting and reporting ... will start to appear from now on, this moment. Then we realized that then it was time to, despite the defeat, rethink strategies, and hence ... keep campaigning' (interview with AS-PTA, 2012).

The disputes over GM crops in this period can be divided into seven episodes. In the first two years, the main target of GM-Free Brazil shifted to the commission of experts. The contours of the CTNBio's powers became clearly defined by the episodes in 2007 and 2008 surrounding the approval of GM corn. Meanwhile, GM-Free Brazil started scaling up by targeting international institutions where there were ongoing negotiations to shape international legal instruments as well as organizing transnational protest events. As years passed and Brazil transformed itself into the world's second producer and exporter of GM crops, GM-Free Brazil shifted its target to civil society and to influencing public opinion on the consequences of their widespread adoption, particularly genetic contamination of corn and increase in pesticide use. New GM products reached the agenda of the CTNBio and Brazil approved its first nationally developed GM seed, a source of pride among the pro-GMO coalition. The advancement of the project of bio-hegemony, however, also caused disputes among its main supporters. Figure 5.2 indicates main events in this period that will be described in more detail below.

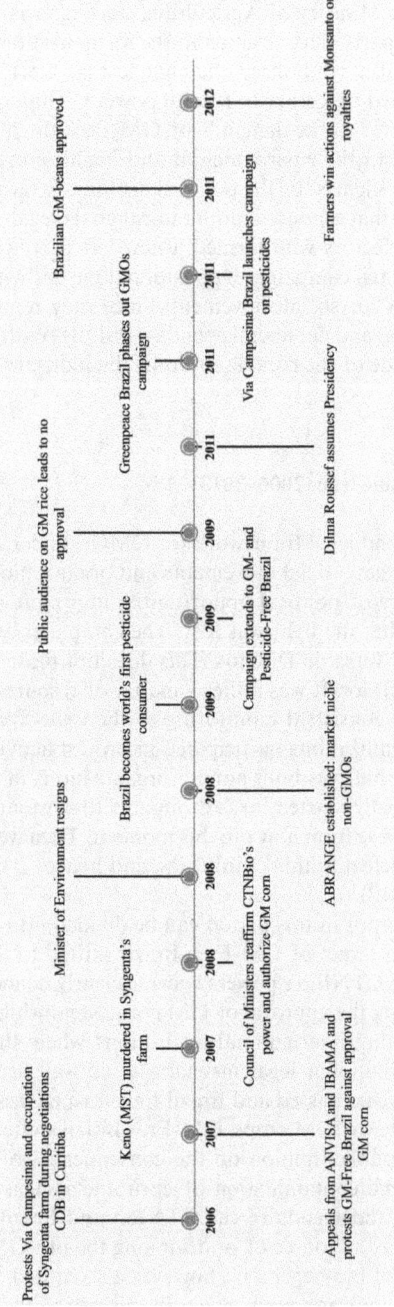

Figure 5.2   Timeline of events in social mobilization against GM crops in Brazil (2006–2012)

Source: Own creation.

Monitoring experts: science and democracy

The context is indeed a new one and it meant, at first, that institutional channels for influencing GM policy were closed. Whereas movements could previously access the parliament, which is an institution officially permeable to the influence of civil society, now decision-making was transferred to a closed commission of scientists. Although not isolated from society, the membership, decision-making criteria and operation implied an attempt to close the public debate and participation in the issue in favour of a technocratic decision-making.

Nevertheless, social movements soon challenged such closure, creating opportunities for participation by means of disruptive strategies and legal mobilization, in which the public attorney became a crucial ally. They knew that a law does not enter into force automatically, and that, without struggles, only the articles of interest for the biotech industry and the *ruralistas* would be implemented.

> [T]he issue of transparency, ... is something that happens today, with all its limitations, but it works not because it is spontaneous, it is written in the law and then it works. It happens because the [social movement] organizations are there ... Despite the fact that it was in the law, we only achieved it by judicial means.
>
> (interview with AS-PTA, 2012)

The opening of the commission to the public was also positive for the media and for firms, which could send their lobbyists to monitor meetings. Still, with the exception of public audiences, activists had no voice: '[W]e went as a listener, not to lose the thread of what was going on' (interview with Greenpeace Brazil, 2012).

In their participation at the CTNBio, activists counteracted the technocratic frame with a democratic one. This was a process of frame bridging, adding the 'democracy' frame to the initial ways to frame GMOs as an issue of environment, health, consumer rights, peasant rights and legality. This new frame entailed ideas such as 'public interest', 'participation', 'transparency' and 'public responsibility', and was used in opposition to practices described as 'closed doors', 'illegality' and perceptions of illegitimacy. This was not due to lack of scientific expertise but it meant, instead, a denial to restrict the debate to those terms. Moreover, avoiding scientific language coincides with the action repertoire of many organizations, which aim at reaching a broader public attention and for which a rights-based discourse has more salience. The activists were stigmatized by the pro-GMO coalition as anti-scientific and obscurantists, relying on the dominant value system according to which science functions as powerful ideology to legitimize politics. However, activists did not evade the scientific debate as many organizations had scientific expertise as well as allies among the experts in the CTNBio. The commission became very polarized following the fault lines in government. The animosity between the two camps demonstrated how science is also contentious: they reflected different epistemologies, one supporting the precautionary principle (interview with Nodari, 2013) and the other defending the principle of substantial equivalence – not as a starting point for risk assessments but as a conclusive assumption of absence of risks. The latter would often accuse Greenpeace of defending foreign interests (interview with Greenpeace Brazil, 2012), a typical trope against transnational environmental organizations by nationalist Brazilians

(Hochstetler and Keck 2007), although they themselves sided with multinational bio-technological corporations.

Activists from GM-Free Brazil understood that science is political and that when applied to biotechnology science has strong ties to economic interests. Thus, they con-tested the idea of disinterested science, on which the GMO policy aspired to gain legitimacy. They repeatedly demanded that the members of the CTNBio sign a declar-ation stating the absence of conflicts of interest, but it was without success. In addition to scientists being funded by biotech industries, activists had doubts about scientists being able to self-regulate their research on biotechnology. Distinguishing biosafety from biotechnology, activists claim that there is a conflict between worldviews, since the former is concerned with assessing risks of the latter, which is always moving ahead. Through the selection of professed biotechnologists (such as all presidents of the CTNBio have shown to be)[16] to lead the biosafety institution, an institutional pol-icy culture is established that rules out the existence of risks and bypasses the objective that created the commission in the first place (interview with AS-PTA, 2012). This attitude can be seen when the president of the CTNBio advanced a proposal to lift requirements for monitoring the effects of GM products after their approval for com-mercialization (Zanatta 2009). Searching for forms of legitimizing the public role of scientists in the regulation of GMOs, GM-Free Brazil expected an institutionalized process of choice from experts to guarantee some representation. This would foster the scientific debate on the issue as representatives of each institution would return with information to their bases and promote a process of reaching a common position so that their vote in the CTNBio reflected not an individual but an institutional pos-ition of a sector from the scientific community. GM-Free Brazil organized an event to mobilize the scientific societies, but this was in vain (interview with AS-PTA, 2012).

Even the SBPC changed its position. During the two mandates from Marco Antônio Raupp (2007–2011), who became minister of science and technology, the SBPC took on a favourable position towards biotechnology, nominating scientists to the CTNBio who voted in block in favour of endorsements. According to a former member of the CTNBio, the ideal of science as a place for open controversy and exchange of arguments could not be further from the reality in the meetings. These were characterized by the disqualification of voices expressing objections. The other myth that had to be debunked was that scientists orient themselves according to the principle of searching for the truth. Any suggestion to demand from the biotech-nological firms further studies or to consider studies that showed evidence of risks related to products approved or under approval by the commission were promptly dis-missed (interview with Kageyama, 2013). These polarized debates dominated the first two years of the CTNBio's functioning, during which no new GM seed was approved. Activists remember that at this moment the technology was 'still' seen as a 'controver-sial issue' and that there was a shared perception of the need to go slowly (interviews with AS-PTA, 2012; Greenpeace, 2012). For the pro-GMO coalition, this meant that CTNBio was not working in a satisfactory manner.

GM corn: no ambiguities left

In 2007, petitions for approval of GM corn dominated the agenda of the CTNBio. This raised the alarm for GM-Free Brazil and brought renewed involvement from peasant movements. Maize raised specific concerns in comparison to soy because of two aspects: maize is the mostly disseminated cash crop for small farmers and there is a high probability of genetic contamination due to its biological characteristics. Although peasants produce soy, it is on a small scale and not so central to subsistence. Maize is identified as the crop for peasant farming in Latin America (interview with MPA, 2012). Activists explain that there is always the danger of contamination of conventional crops by genetically modified varieties. However, in the case of soy, this only occurs through physical means within close proximity. By contrast, the maize plant cross-fertilizes, which can lead to genetic flux between the GM variety and the conventional one across large distances. The combination of these two factors threatens farmers' control over their processes of seed selection and farmers' rights over seed multiplication because intellectual property rights become an unavoidable problem since all GM seeds are patented.

The discussion about farmers' rights became central in the debate over GM crops in Brazil. Because of the tangibility of this threat, it was easier to make the debate about GM crops with small farmers (interview with MPA, 2012). Due to their accumulated experience with corn, they knew that maize crops cross-fertilize and that the same characteristic that empowered them as agents of genetic selection was now threatening them. The traditional practices of saving and exchanging seeds provided the basis for the discursive articulation of varied issues such as agrobiodiversity, peasant culture and farmers' rights over seeds. The master frame of food sovereignty took on its strongest expression, reinforcing the interconnections of these issues: as small farmers and peasants were the ones who historically selected seeds, according to cultural and other needs in addition to economic productivity, they promoted biodiversity of corn.

In addition, farmers' rights of saving a part of the harvest as seeds for the next sowing season was also under threat. This practice is part of the cultural identity of farmers who were alarmed by rumours of the 'terminator seeds'[17] and the prospect of planting a year, harvesting and, when planting again, the seed would not germinate. GM corn was composed of a genetic modification inserted on a hybrid seed that is designed to be sowed only once, thereby forcing farmers to go to the seed market every season. This provided the entry point for the discussion of the corporate food regime because it puts an end to farmers' autonomy as it makes them dependent on the seed market under the control of a few firms. The existence of different realities in the country helped in the exchange of experiences. Farmers from the MPA in the regions of the south-east and south could tell others from the north-east, north and centre-west about their experience of having converted to hybrid seeds and concomitantly having the diversity of their crops reduced to what was offered on the market (interview with MPA, 2012).

Last by not least, farmers lose control over their seed selection when a transgene and a patented gene enter into the genetic mix of their corn varieties. The issue of contamination by GM corn illustrates the deep association between biotechnology and intellectual property rights (Góngora-Mera and Motta 2014; Kloppenburg 2004; Pechlaner and Otero 2008; Rauchecker 2013). GM-Free Brazil activists frame the double nature of the problem: a biological nature due to the insertion of gene external to the corn species that changes the farmer's practices of selecting seeds, and a socioeconomic one, as genetic modification is a proprietary technology legally protected by a patent. It falls under the *Law of Industrial Property* that incorporates the TRIPS obligations. Hybrid seeds are regulated by the *Seed Law*, which follows the UPOV

1978 Convention. The latter allows for self-reproduction of seeds and exchange or sale among family farmers, practices that are forbidden with the advent of patents over seeds (Kloppenburg 2004; Reis 2012). In sum, for all these reasons, GM corn mobilized most organizations in the campaign.

The approval of GM corn

As the CTNBio was not meeting the expectations of the pro-GMO coalition, the latter decided, once again, to change the law. The executive power issued Provisional Measure 327, amending the *Biosafety Law* to change the voting majority at the CTNBio from two-thirds to an absolute majority. The protests of GM-Free Brazil were disregarded and, in 2007, the National Congress ratified the measure. It was only with this amendment that the institutional power of the pro-GMO coalition became hegemonic in the expert commission and the official policy became clearly favourable to GM crops. This shows that the networks of the bio-hegemonic project had permeated not only the executive branch but also the legislative, so much to the extent that it actively provoked the legislative to change the law.[18] It marked a turning point in the standstill of GMO approval: in September, three varieties of GM corn were approved for commercial use.

GM-Free Brazil reacted with various strategies. Pregnant women from urban and agrarian movements from the Movement of Peasant Women (MMC) entered the room during the CTNBio meeting holding posters with the slogan 'my son is not a guinea pig' (Pereira and Fernandes 2007). Activists collected signatures for a petition sent to the president and proposed an injunction to suspend the decision. It had a partial victory, with the prohibition of the cultivation of the product in the north-east region until environmental assessments can be conducted for the specific biomes of the region. The issue of coexistence became the object of a second judicial action (interview with AS-PTA, 2012).

The approval of GM corn drew the ANVISA and IBAMA into the battle. Using their prerogatives foreseen in the *Biosafety Law*, each agency appealed against the approval of GM corn, challenging its scientific basis. This provoked the positioning of the CNBS for the first time. It met in February 2008 to deliberate on the appeals, but ignored the contents and only addressed the institutional issue. Relying on the opinion issued by the general attorney in 2002 in the context of the judicial moratorium on GM soy, the CNBS settled the issue as follows: the CTNBio has the final word on issues relating to the biosafety of GMOs. So the three varieties of corn were approved without considering the objections that motivated the appeals from the IBAMA and ANVISA. The CNBS issued two recommendations: first, that the CTNBio consider scientific studies conducted by third parties in order not to rely on studies made by the proponent firms, justifying when this is not followed; and second, the creation of an interministerial working group to conduct studies of the medium- and long-term impacts of GMOs and their effects on health and environment (CNBS 2008a, 2008b). These were never followed.

The approval of corn was not a technical but a political decision (interview with former official IDEC, 2013), which gave GM-Free Brazil and their allies in the government a clear message: do not use the legal room for appealing against the CNTBio. The extensive use of legal mobilization in the disputes over the biotech neoliberal food regime in Brazil meant, if not a substantial victory, at least a procedural gain for

GM-Free Brazil, and ensured a minimum room for the observance of legal requirements. From the point of view of the pro-GMO coalition, it meant successively delay-
ing their script. But this does not mean that they made less use of legal means.[19] The
organizations from GM-Free Brazil interpreted the approval of GM corn as the
second big loss. It marked the speeding up of approvals, making it more difficult for
movements to monitor. GM-Free Brazil understood that the spaces for their influence
in the executive power were closed. It meant, for many, a final definition of the policy
regarding GM crops, that is to say, one of fostering it by letting the CTNBio work as
it pleased. In short, in the politics of GM crops health and environmental state bodies
had no voice; the government had silenced them as a consequence of the decision to
support all demands coming from agribusiness.

Monitoring the National Congress

This was a moment of critical assessment for the GM-Free Brazil campaign. For
Greenpeace, what had been a fundamental topic in the environmental agenda started
to lose relevance. The MST understood that the CTNBio was no longer an arena
for their mobilization efforts. The campaign kept monitoring the National Congress
given the reoccurring attempts of the pro-GMO coalition to dismantle any legal provision perceived as an obstacle to advance agribusiness. This demonstrates that the
legal framework in force resulted from a series of compromises with social movements
and did not fully correspond to the interests of the proponents of the technology. It
also illustrates how law is a contentious field. Among bill projects developed by the
bancada ruralista, two raised special concerns for activists.

In 2008, Deputy Luis Carlos Heinze (PP-RS) introduced a project bill to end labelling requirements for products containing GMOs. The then Deputy Kátia Abreu –
later senator, president of the CNA and current minister of agriculture – created a
project intended to remove from the *Biosafety Law* an article that prohibited the use of
'terminator seeds', which was undertaken as Bill Project 268/2007 by Deputy Eduardo
Sciarra (DEM-PR). In 2009, Deputy Cândido Vaccarezza (PT-SP), who participated
in the *trenzinho da Monsanto*, combined both goals into one project.[20] GM-Free
Brazil organized protests and motivated their allies in the parliament not to let the
bill advance.

Vaccarezza was by then the leader of the federal government in the Chamber of
Deputies, which not only shows how the pro-GMO coalition found allies in the PT
but also that precisely those who converted to the agenda of agribusiness reached the
highest positions in the party and the governmental hierarchies. Not by chance, this
period also witnessed an offensive of the *bancada ruralista* in many other issues, such
as the postponement of voting a law on the enforcement of penalties against the use of
slave labour and the development of a law project to amend the *Forest Law*, introdu-
cing flexibility in defining environmental crimes and with the provision of amnesty for
past violations due to agrarian activities. The 'terminator project' remained a threat in
the Brazilian legislative.[21]

Scaling up: transnational protest events and international conferences

Meanwhile, the campaign expanded its actions to other scales of contention. In 2006,
Brazil was going to host the eighth meeting of the Conference of the Parties to the

Convention on Biological Diversity (COP 8). This would be preceded by the third Conference of the Parties, which would serve as the meeting of the Parties to the Biosafety Protocol (COP-MOP 3), the *Cartagena Protocol*. These meetings provided a good opportunity for protest events, taking place in Curitiba, Paraná. The governor Roberto Requião, an outright opponent of GM crops, had been newly re-elected, the peasant movement had strong bases there with experience and repertoires from previous cycles of protest, and Terra de Direitos, with expertise in national and international law, had its headquarters in Curitiba. This intersection of factors formed a favourable context for the scaling up of GM-Free Brazil to the international level.

During the two years leading to the conference, the organizations started their preparatory process. Terra de Direitos became familiarized with the topics and procedures of the international meetings and acted as proponents of the grass-roots movements' goals, publishing periodical bulletins, organizing workshops and meetings, and representing civil society in the governmental consultations to define the Brazilian position (interview with Terra de Direitos, 2013).

The minister of environment obtained some concessions from the president in advancing the biosafety agenda in Brazil. This enabled the country to aspire to a leading role in the negotiations, which was compatible to its position as the host of the conference. In 2006, the president issued a decree on the implementation of the protocol, which had been ratified by the National Congress in 2003, and met with the ministers of environment and agriculture to agree on the position regarding labelling rules in the negotiations of the protocol. The position of the Ministry of the Environment prevailed, with Brazil defending a clear labelling for international trade with GMOs. This would imply that Brazilian exporters would have to undertake segregation and testing of their products – which was never implemented (Lopes and Tortato 2006).

On the verge of the COP-MOP to the *Cartagena Protocol*, which was going to start on 13 March, about 2,000 women activists from Via Campesina, affiliated with the MMC and MST, decided that their political activity on International Women's Day (8 March) would be a protest against GMOs. They destroyed a laboratory of a pulp and paper industry that had been conducting experiments with transgenic eucalyptus (interview with MST, 2013). By choosing a day with international meaning for the women's movement and demands for rights, and making use of violence in the proximity of an important and thematically related conference, the protest received media attention both nationally and internationally. The protest activities from the peasant movement did not stop there.

During the COP, Via Campesina mobilized around 3,000 peasants, who camped around Curitiba and organized the Global Forum of Civil Society (Tortato 2006). On 14 March, Via Campesina chose a transnational target for their protest, the multinational seed corporation Syngenta. A week before, the IBAMA in Paraná had denounced that Syngenta had been ignoring the national legislation for the experimental cultivation of GM corn in the buffer zone of a national conservation park (Baptista and Tortato 2006). Via Campesina occupied the farm. After being evicted from the farm, they camped on the roadside. A number of aggressive acts followed, among which, they were sprayed by air with pesticides (interview with MST, 2013). Via Campesina reoccupied Syngenta's farm (Via Campesina 2006). A long conflict ensued, which culminated on 21 October 2007 with the murder of a member from Via Campesina, Walmir Mota de Oliveira, known as Keno. For activists, it was a communicative act: it was perpetrated with the intention to exemplify, to show others what would happen if activists continued mobilization.

The murder raised international attention. Terra de Direitos became involved with a judicial action to investigate the murder and Via Campesina used international venues to widen the repercussion of the event. The next COP of the Convention of Biological Diversity provided a good opportunity. It took place in Bonn, Germany, in May 2008. The MST sent representatives, who handed in a report on Keno to the Secretariat of the CDB. There was a lot of international pressure and events organized in front of Syngenta's headquarters. In the end, Syngenta decided to withdraw from the area and the government of Paraná planned to establish a centre for teaching and research on agroecology named after Keno (interview with MST, 2013).[22]

Regarding the COP-MOP as such, it was the first time that the MST was part of the Brazilian diplomatic delegation. This signalled the opening of an institutional space until then very hesitant to movement participation. However, this openness was not translated into equal participation. When it came to defining the Brazilian position in international negotiations, the activists recognized that they were at a very disadvantaged position vis-à-vis the number and preparedness of the private sector: 'the Brazilian delegation reached ... 300 members, 90 per cent were businessmen' (interview with MST, 2013). This asymmetry was reflected in the real possibilities of influencing the decision-making. The physical presence of the activists served more as a learning opportunity to understand how it actually worked than to participate and be heard in the formation of a common position. This cannot only be attributed to the differences in number and know-how on diplomatic issues. Above all, it reflected the governmental position of backing, in the international arena, the demands of the agribusiness lobby, which was facilitated by Marina Silva's resignation from the office as minister of environment. She resigned, following the approval of GM corn, precisely when the Brazilian delegation was expecting her arrival to head the negotiations.

This was also the occasion for GM-Free Brazil to attempt to achieve a 'boomerang effect' (Keck and Sikkink 1998) in the form of sanctions on the part of the CDB. They denounced that the authorization of GM crops in Brazil did not comply with its obligations under the *Cartagena Protocol* (Salazar 2011), but they found no leverage on the international level. Nevertheless, morally, it did show Brazilian diplomacy in a poor light (interview with MST, 2013).

Facing contamination

As stated before, activists interpreted the approval of GM corn not as an end of their activities but as the beginning of hard work. In 2008, GM corn already reached the fields and the first harvest was expected for the first months of 2009. The first issue GM-Free Brazil addressed was the imminent contamination by GM corn. In a wide attempt, they organized a seminar in Curitiba in order to revive the original ideas of the campaign. This brought the campaign closer to the agrarian reality. Small farmers became the main targets of information campaigns, aimed at providing them the skills to detect contamination. Among the instruments, strip tests to make seed contamination visible were distributed to farmers or to farmers' unions. Farmers promptly reacted; they started sending pictures from experimental fields as well as denounced irregularities (interview with AS-PTA, 2012).

The idea of giving farmers the means to protect themselves from GM corn had cultural resonance. The impact of the messages from GM-Free Brazil was a combined product of the empirical credibility of the spreading of GM corn, which was publicized; the credibility of the frame articulators, who had long been working with small farmers; and their frame consistency, as concerns about contamination were not something new but part of the promotion of agroecology in the work of the AS-PTA and the GM-Free Brazil campaign. They designed the initiatives 'seed guardians' and 'biodiversity guardians' to promote political recognition of the act of saving and protecting seeds by organizing meetings and giving farmers certificates. These activities resonated well with the widespread understanding among small farmers and peasants that seeds are 'public goods' (interview with AS-PTA, 2012). The state of Paraná took important actions to address the problem of contamination: it worked in collaboration with the MST in identifying contamination of the fields (interview with MST, 2013) and sent a document reporting on the inefficiency of the coexistence rules set by the CTNBio, in turn demanding new rules.

Meanwhile, the majority of corn farmers who planted hybrid seeds sold at the commercial market fell victim to the market strategies of the biotech industry. There was less and less supply of non-GM corn varieties. The AS-PTA estimated, based on data from the Ministry of Agriculture, that by 2010 more than three-quarters of new seed releases were genetically modified. Relying on industry data, they calculated that 20 per cent of corn farmers were using creole seeds and 80 per cent were relying on commercial seeds. The latter are used to buy seeds every year, and due to the marketing decisions of seed industries of only releasing GM varieties, these farmers had no choice but to buy those. There have only been some marginal initiatives by regional agencies of multiplying conventional seeds (interview with AS-PTA, 2012). Indeed, whereas GM soy took about ten years to cross the threshold of 70 per cent of cropped area, GM corn did it in only three years.

Targeting pesticides: uniting efforts or diverting attention from GMOs?

Since 2009, Brazil has been the world's top consumer of pesticides (Carneiro *et al.* 2012). Different events confirmed activists' warnings that GM seeds were not the environmental friendly and efficient technology proclaimed by their proponents. Problems started when other plants became resistant to glyphosate – the pesticide used in combination with RR seeds – and transformed into pests for the crop (the super weeds), or insects developed resistance to the genetic modification (Bt) and became pests (Cintra 2013; Globo Rural 2013). A future risk resulted in very tangible damage that affected many farmers. Indeed, there is a visual component in the power of GM seeds to demonstrate the efficiency of glyphosate: GM soy fields are described by their owners as 'a beauty', completely 'clean', without any plagues; by contrast, critics call them 'green deserts'. However, even enthusiasts of GMOs recognized the limits of the technology when resistance developed ('Dez Anos de Transgênicos No Brasil – Caminhos Da Reportagem' 2014). In Rio Grande do Sul, with a longer history of GM soy, farmers started reacting by using more pesticide and more toxic products, such as 2,4-D, which was one of the active ingredients of Agent Orange used as a chemical weapon in the Vietnam War, and its contamination of the environment still victimizes many. They had to apply it before sowing soy and this increased its volatility: it contaminated the neighbouring lands, which also increased conflict (interview with Greenpeace Brazil, 2012).

This also became a topic of the seminar in Curitiba in 2009. Since then, the campaign modified its name to For an Ecological, GM- and Pesticide-Free Brazil. Activists perceived many gains in extending the scope to the issue of pesticides. There is a wide degree of consensus in Brazil that pesticides are bad for health, thereby accumulating experience on the topic. The bridge to health guaranteed many allies in the scientific community, among which the prominent National Cancer Institute (INCA) and the Brazilian Association of Collective Health (ABRASCO), which launched a dossier on pesticides in 2012 (Carneiro *et al.* 2012), and two more in the next years. For all these reasons, activists contend that pesticides have a stronger resonance for consumers than GMOs.

The impacts of pesticides on the public health of rural workers and communities are much more acute. Small farmers felt it on their bodies and in their pockets:

> Many families have today two main problems that have facilitated it for us to make the debate … First, the contamination itself, and it reflects on the health of peasant families … and there is a perception … that poison can damage and must be reduced not to worsen the illness. [T]he other is the economic issue, because it is very expensive today to buy the [technological] package.
>
> (interview with MPA, 2012)

In 2011, the organizations from Via Campesina Brazil launched the Permanent Campaign against Pesticides and For Life, which received support from various organizations. They determined that it is easier to campaign against pesticides than against GMOs for three reasons. First, because they considered that they lost the fight of GMOs:

> So, why another campaign? Our evaluation was … [that], we were hit … by a government identified with popular movements, on which we posed our expectations that it would carry forward the struggles from social movements. We ended up with the contrary of all we expected.
>
> (interview with MPA, 2012)

The second reason was about the ease of promoting the debate on issues among peasants. Pesticides were a good strategy to make the debate larger concerning the agrarian model (interview with MPA, 2012). Indeed, with the rise of pest resistance and, correspondingly, of pesticide use, even the first economic benefits of GM seeds experimented by small farmers were reversed:

> It increased productivity at first, but today to keep it, you have to spend more. I went to many seminars, the peasants speaking like that: last year I spent this to produce that, this year, I will have to spend more. So, they have real data. Today it is more perceptible. At first it was very difficult. That is why I think we were defeated in transgenics, at that time.
>
> (interview with MPA, 2012)

Some anti-GMO activists have criticized the campaign from Via Campesina due to the danger of such a shift in focus; the representative of MPA maintains that the frame bridging between GM crops and pesticides will hold:

> Because they are inseparable: there is no manner to make the debate on pesticides without debating about transgenics ... the firms who are petitioning for transgenics are also the biggest producers of pesticides ... and the most part of approved seeds have a direct link to pesticides. The data shows.
>
> (interview with MPA, 2012)

Third, there were many more openings in mass media due to two aspects. Pesticides are associated with cancer, an illness with wide public resonance. In addition, activists perceived a converging of interests that brought together mainstream media, biotech firms and big farmers. The difference of media attention and sympathy to the pesticide issue, the spokesperson from the MPA explains, lies in the suggested alternatives, such as organic foods, which can be accommodated under the logics of agribusiness (large scale, with lower cost once an area has been brought under ecological equilibrium, and higher profit margins). He contrasts this with the case of GMOs: '[W]ell, not take GMOs but what could be used that would benefit them? Nothing. There is not alternative that somehow benefits them' (interview with MPA, 2012). By relying, once more, on the master frame of food sovereignty, the movement worked on the problem and the solutions to it. They pursued the debate that it was not enough to produce without pesticides but also depended on how the production is made, on which land, what size of property, and with what type of relation to the environment.

Everyday food on Brazilian tables: unexpected allies against GM rice and a new national feat, GM beans

The next food crops to land on the agenda of the CTNBio were rice and beans, the two main ingredients of the Brazilian food culture. Different from soy, a crop that in Brazil remains distant from the food plate, and from corn, from which the majority is destined for animal feeding, these two food crops are mainly for human consumers. Despite that, the eminence of the approval of GM rice and beans did not cause much social mobilization among consumers, even though they found some resonance in the media (interview with AS-PTA, 2012). The two histories also differ.

There had been mobilization against the approval of GM rice in 2009. Greenpeace had a prominent role following the decision from Greenpeace International to launch a global campaign for impeding the world approval of GM rice. Together with organizations from GM-Free Brazil, they organized events to raise awareness. The GM rice under petition was modified to be tolerant to a pesticide, glufosinate, considered to be more toxic than glyphosate, and this raised concerns on an already controversial issue, pesticide use. Greenpeace advanced the argument of the risks of genetic flux with wild species, thus a threat to biodiversity. The MST also mobilized, being the biggest producer of ecologic rice in Brazil. The CTNBio organized a public hearing about GM rice. This time, social movements encountered a window of opportunity: associations of farmers, potential beneficiaries of GM crops, and Embrapa, a major proponent of biotechnology, contradicted their, until then, consistent pattern of defending GMOs (Motta 2015; Zanatta 2009). In the end, the firm that had filed the petition for GM rice, Bayer, withdrew it. The activists were not naïve to interpret this fact as a positive outcome of their campaign. Rather, they mention the role of unexpected and powerful allies because of market and agronomic reasons.

By contrast, the position of Embrapa, when its scientists announced the launching of GM beans in 2011, was professedly advocatory. The pro-GMO coalition promoted it as a national feat. This resonated well with a nationalist political culture such as Brazil. There was also a technical difference in the genetic modification: beans were not modified to resist pesticides but to resist a plague. Considering the context of a stronger sensitization to the pesticide issue, genetic modifications not associated with pesticides could be met with less resistance, as was the case (interview with AS-PTA, 2012). In addition, GM beans arrived when the GM-Free Brazil campaign suffered a process of demobilization. Greenpeace Brazil had already phased out the campaign against GM crops as part of the reorganization of its efforts to focus on climate change (interview with Greenpeace Brazil, 2012). Also the peasant movement concentrated on the national pesticide campaign, owing to the considerations that the issue of GM crops could no longer motivate mobilization, as explained by one spokesperson:

> I think it has to do with the logic of *fait accompli*, that feeling of defeat. I think this has an impact on the self-esteem of the struggle, in mobilization capacity, the ability to fight ... What I mean: good, in the case of soybeans, it passed; corn, passed; rice, well, there I think we got it ... Then, the beans, and with Embrapa proposing even. So that was sort of given, as fact.
>
> (interview with MPA, 2012)

On 16 September 2011, GM beans were approved (Yano 2009). The National Council on Food Security and Nutrition (CONSEA), a consultative body composed of representatives from civil society and the government, reacted demanding the president apply the precautionary principle, but it was ignored. Activists changed their strategies, foregoing the use of confrontational and disruptive strategies. While many shared the feeling of having been defeated, others believed that the consequences of the adoption of GM crops would prove the technology wrong. The organizations that keep GM-Free Brazil active have many campaign fronts, including legal activism, monitoring of the CTNBio and the parliament, and communication activities.

Emerging disputes in the bio-hegemony project

The approval of GM beans already took place during the government of President Dilma Rousseff (PT), the candidate supported by Lula, which meant continuity with his government. There were even clearer signs of the strengthening of the power of agribusiness in the governmental coalition as Senator Kátia Abreu officially entered the government by changing her political affiliation to the PMDB. In 2015, she was nominated minister of agriculture under Rousseff's second term.

As Brazil converted more and more agrarian fields to biotechnology, three issues of dispute emerged inside the bio-hegemony project: the enforcement of intellectual property rights over seeds, the possibility of making Brazil also a major world supplier of non-GM seeds, and the uncontrolled pest and plague resistance. In Brazil, Monsanto has the patent on the RR soy and has developed a system to collect royalties on its seeds. Without challenging the patent itself, the FETAG-RS took legal action against Monsanto's practices of charging royalties over harvest at the moment of commercialization.[23] In another action, the farmers' union from Mato Grosso, Femato, together with the Association of Soy Producers (APROSOJA), challenged Monsanto in 2010 for unjustifiably collecting royalties after the patent expired while demanding the decrease of the amount paid. Having received favourable decisions from regional courts, and, with the appeal from Monsanto, the case reached the Supreme Court in 2012. Meanwhile, Monsanto tried to settle the issue with farmers by offering them, in individual contracts, royalty-free new soy in exchange for their removal of the suits (Kaskey 2013). Farmers' associations instructed their members not to sign it while the Supreme Court decided on the case. The Supreme Court ruled in May 2013 that Monsanto's 20-year-old patent had expired, making it public domain. The company was ordered to return to farmers double the amount charged; it appealed.

The other division regards the coexistence of non-GM and GM crops inside agri-business, and of state support for both cases. Since 2008, grain processors and traders created the Brazilian Association of Non-Genetically Modified Grain Producers (ABRANGE), looking for a market niche. Whereas Embrapa played a role in the development of conventional seeds, the Ministry of Agriculture maintained a clear position in all pro-agrobiotechnology forums. However, the lack of recognition by the Ministry of Agriculture is detrimental to expansion of a non-GMO market niche and ABRANGE has been struggling for political space.

Since 2012, an enormous economic loss is estimated due to a pest of the earworm caterpillar, *Helicoverpa zea*, which is decimating GM cotton, soy and corn crops. The spreading of GM Bt maize has been blamed for it as it killed off natural enemies of *Helicoverpa armigera*. In some places, a state of emergency was declared and the Ministry of Agriculture and the *bancada ruralista* used means to legalize the import of highly toxic pesticides not approved in Brazil. This included the corrupting of public officials responsible for pesticide approvals and, for those who denounced 'irregularities', the response was nothing short of dismissal.[24] In the meantime, in 2013, Monsanto started to market its new GM technology as a solution to the plague: again, it genetically modified to tolerate pesticides. Dow AgroSciences launched GM seeds tolerant to a much more toxic substance, 2,4-D, claiming that it is safe (Dow AgroSciences 2012).

Notes

1 She started her activism in the Amazonian state of Acre in the early 1980s, among the Catholic Left and rubber tappers. She worked together with Chico Mendes to establish the branch office of Central Workers' Union, forming the bases for an environmental movement in the Amazon. In 1988, running for the PT, she was elected to the local legislative and reached the federal parliament as a senator in 1994 – since then she has been re-elected. From 2003 to 2008, she served as environmental minister in Lula's government. In 2009, she left the PT and started a political project to be the president of Brazil.

2 In June 2004, China rejected four shipments of soy for the presence of fungicides, a product used to treat seeds before sowing, which substantiates the hypothesis advanced by GM-Free Brazil.

3   The Minister Roberto Rodrigues, said in a statement to the press after the release, he spoke of the 8% of the national harvest, corresponding to 4 million tons. Previously, some government officials from various ministries had spoken of 30% of the national harvest, or 15 million tons. Other had estimated 70% of the RS harvest, 30% of the Paraná harvest and 15% of MS harvest, resulting in 9.6 million tons.

(Fernandes 2005, 4)

4 Many activists in Brazil have used this expression. Already in 1978, environmentalists referred to it to criticize the military government decision to build an airport in a forested area (Hochstetler and Keck 2007, 78).

5 The Socialism and Freedom Party (PSOL) was founded in 2004 from leftist dissenters from the PT.

6 In 2003, they numbered 103 deputies and five senators (Cesarino 2006, 83; Taglialegna 2005, 43, 44) and managed to appoint two of the five rapporteurs of the bill.

7 The exemplary case is Cândido Vaccarezza (SP). His power increased to the extent of becoming the leader of the national government in the Chamber of Deputies from 2009 to 2012. Other PT politicians participated in the trip: Fernando Ferro (PE), Paulo Pimenta (RS), Zé Geraldo (PA) and Josias Gomes (BA).

8 Brazil appears to be unique in this respect, as other regulatory systems differentiate sharply between red (medical) use and green (agricultural) use of biotechnology. Surveys have shown that the individuals also differentiate their opinions according to this division (Bauer 2005).

9 Lisboa is one of the founding members of Greenpeace Brazil and was its campaigner on GMOs. She was invited to head one of the secretaries of the Ministry of Environment and has represented civil society in the CTNBio.

10 The legislative power was and would remain active in defining the policy on GMOs. Camara (2011) analysed parliamentary discourses from 2003 to 2008 and found 887 discourses on GMOs.

11 The social security system guarantees the right of every rural worker to receive a pension in the value of a minimum wage. As 98 per cent of rural workers receive it, the increase in the minimum wage has had an impact in reducing rural poverty (Lavinas 2012).

12 Her work is based on the media clipping from the GM-Free Brazil campaign (*Boletim*) and the media clipping from ANBIO, the organization of biotech firms, during 2003 and 2005.

13 Guivant (2006) also noted the public relations strategy of Monsanto. It hired the research institute Ibope to conduct a survey in 2003 about a publicity campaign broadcasted nationally between 8 and 28 December 2003 targeted at persuading watchers of the benefits of GMOs.

14 Nobre (2013) creates the concept in reference to the PMDB, for being the protagonist of a movement to tame demands for democratization, containing it under its spheres of influence, namely an elitist political system. *Pemedebismo* channelled political conflict to the parliament, where it created a system of dispute resolution based on backroom and corridors politics and on the creation of a big coalition in which

all parties had a share of power. This was disseminated as a necessary condition for exercising power in Brazil: 'governability'. Until then, the PT had rejected such practices.

15 Singer (2012) conceptualizes these changes in terms of the two souls from the PT. The first soul is the foundational one, based on the workers' class identity and a radical socialist programme of social change. The second soul was born in the electoral campaign of 2002 and has become dominant ever since, characterized by its accommodation to the national political culture, marked by conciliation among elites and ambiguity.

16 Walter Colli, who presided over the commission for two mandates, has repeatedly voiced that there would never be risks from GMOs (Motta 2013b).

17 Terminator or 'suicide seeds' are GM seeds based on genetic use restriction technologies (GURT) that prevent the use of second generation seeds by making them sterile, which forces the farmer to buy seeds every season.

18 I am grateful to Gabriel Fernandes for these remarks.

19 Bissoli (2012) found 28 legal actions on GMOs in Brazil from 1998 to 2012, used by both sides.

20 The deputy stays loyal to this project (Vaccarezza 2014).

21 In November 2013, the bill project from Deputy Sciarra was in the parliament (Thuswohl 2013). Brazil appeared in international news as threatening to suspend the global moratoria in this GM technology (Watts 2013).

22 This took place in a moment of rising violence and of criminalization of social movements by state authorities in Rio Grande do Sul. In 2008, the Public Prosecutor's Office requested authorizations to evict 160 families from the MST – which were conducted by the military police with violence – and commanded investigations to charge its members with organized crime.

23 On 14 June 2012, the judiciary ordered the suspension of the collection of royalties over harvest with the argument that GM soy was not protected under the intellectual property rights law as a patent but under the *Plant Variety Protection Act*, which gives farmers the right to reserve for their own use and to sell the products without royalties. The Superior Court of Justice decided that this action would be nationally binding for Monsanto.

24 This was the case of the head of the Toxicological Division from ANVISA, Luiz Claudio Meirelles, who was dismissed in 2012 (Idec 2012).

Chapter 6
Many tales of GMOs: a comparison

The contrast between the stories from Argentina and Brazil has provided some insights of possible explanatory factors for the different trajectories as well as for the changes in them. The previous chapters have followed an inductive approach, constructing a historical narrative, in which the empirical data was interpreted with the aid of concepts presented in the analytical framework. This chapter aims at summarizing the research findings by highlighting the key explanations for each outcome. It will first distinguish between the four outcomes of the disputes over GMOs. Then it addresses the research questions by identifying the three main analytical factors that explain the variations in the outcomes, as well as the role of context conditions. This leads to the question regarding how the structural location of these countries in global commodity chains affect the structure of opportunities for social movements fighting GM crops. These findings are then summarized in an explanatory model.

The different outcomes: bio-hegemony, bio-hegemony with mobilization, controversy and dominance

The first task is to describe the different outcomes corresponding to bio-hegemony, bio-hegemony with mobilization, controversy over GMOs and dominance of GMOs. Each is defined according to the relative approximation to the ideal types of outcomes in a continuum between, on the one side, the construction of bio-hegemony in terms of material, institutional and discursive powers, and, on the other side, the constitution of GMOs as a topic of public debate and political controversy. After describing each of them, I will then clarify why dominance of GMOs cannot be considered to be the same as bio-hegemony.

Bio-hegemony in Argentina

The first outcome described was bio-hegemony in Argentina. It resulted from a solid process of constructing biotechnology as meaning the general interests of Argentinean society. This included coordinated efforts to build material, institutional and discursive powers. Materially, GM crops increasingly dominated lands and research in plant breeding; they brought important state revenues in taxes and as foreign currency due to exports. INTA and scientists became promoters of the new technology. Biotechnologists dominated the expert commission with a policy culture of non-regulation. This guaranteed a fast-track approval system for GMOs. Institutionally, the decision on the new technology was concentrated in the hands of agrarian authorities under a policy framing of innovation, which promoted a positive public meaning of GM seeds. Discursively, the pro-GMO coalition built an extensive network within the civil society, including mediated discourse, spreading their preferred meanings about the processes of transformation under way, such as transgenic revolution, knowledge society, modernization and progress through science and innovation. The legitimation of the biotech food regime was thus the result of the construction of such alliances and discourses, always avoiding an open debate, the exposition to contradictory argumentations and the questioning of any of its components. At this time, there were some actors reacting to the agrarian change in the countryside, and also social movement organizations in the national capital contesting the approval of GM seeds. But their actions did not reach the wider public debate and found no influential ally to leverage their influence and to make the approval of GMOs a controversial issue in Argentina.

Bio-hegemony with mobilization in Argentina

After ten years of adoption of GM seeds, when the biotech food regime was installed, social movements had built an extensive network for mobilization. If alliances between NGOs proved difficult at the beginning, the strengthening of grass-roots mobilization created more opportunities for new coalitions. The (unsuccessful) urban campaigns were replaced by vivid protests in the suburban and rural contexts. GM seeds were rarely identified as an immediate concern and rather as a cause of their problems such as forced evictions, diseases and environmental contamination; these linkages were objects of interactive framing processes. New political actors such as peasants, indigenous peasants, mothers and sprayed people became increasingly organized. These were important outcomes of social movements.

However, when social mobilization against the socio-environmental effects of the agrarian model based on GM seeds emerged, its hegemony was solid enough to silence or ignore claims. There were no openings in the political landscape, no institutional partner to respond to claims against forced evictions and pesticide contamination, and no media space that would cover their claims. Therefore, even if in a second moment there was an increasing level of social mobilization against the socio-environmental effects of the model, it did not succeed in reaching the wider public sphere and making GM crops controversial. Such conflicts, when they went beyond affected people and achieved some level of attention, remained regional in their scope. The only moment when the agrarian model became an issue in the national public debate and politics, was a distributive dispute between the state and agrarian producers, which did not question the model or make it controversial. In short, since the beginning, the pro-GMO coalition led the path in Argentina, with their strategies of using material, institutional and discursive powers, including the creation of a shortcut to bypass public debate. The hegemony of the biotech food regime was not undisputed as activists increasingly contested the legitimacy of the agrarian model and started to gain attention as well as achieve some legal changes by targeting ecological and social impacts. Therefore, this last phase must be described as involving a mechanism of mobilization. However, such disputes did not reach the wider society and could not challenge it. Argentina remained an example of bio-hegemony, albeit with increasing mobilization.

Controversy over GMOs in Brazil

The third outcome narrated was the controversy over GMOs in the history of the technology in Brazil. It resulted from a process of mobilization from social movements and organizations from the civil society. They acted early, before the approval of GM soy. Their actions reached the mass media and found openings in the state, allies in the political landscape, and scientific organizations. The mobilization was consolidated in a campaign against GMOs and a wide coalition for a precautionary approach. The result was the construction of GMOs as a contentious technology, making it an issue and bringing media attention to it. They also succeeded in influencing the policy, as their legal injunction established a moratorium.

At the same time, the actors' network in favour of the approval of GM seeds found obstacles in building bio-hegemony. The regulatory system was not operative as its constitutionality and legality were constantly challenged in the judiciary. There was already a cooperation agreement between Monsanto and Embrapa in Brazil, but it was illegal. The material power from the pro-GMO coalition could not express itself due to the absence of the other types of power, institutional and discursive. Therefore, the bio-hegemonic network in Brazil would rely extensively on illegal means. In regards to institutional power, there were long disputes among state bodies concerning the jurisdiction over GMO policy, with the Ministry of Environment claiming a final saying over the environmental release of GM seeds. There were politicians, bureaucrats and scientific organizations defending precaution. All these positions were portrayed in the mass media, which covered it as a controversy with a plurality of frames, voices and opinions, showing that there was not hegemonic discursive power on the issue.

Dominance of GMOs in Brazil

The initial success of social mobilization in Brazil did not last forever. Having met with such unexpectedly contentious Brazilians, the pro-GMO actors' network reacted and coordinated its efforts to ensure material, institutional and discursive powers as well as to break the coalition of social movements against GM crops and demobilize them. They have increased their material power by illegally expanding the agrarian production of GMOs and concentrating on seed supply, thus creating a situation of *fait accompli*, and as scientific organizations changed their position to openly defend biotech crops. In order to legalize this situation and create a favourable and operative regulatory system, the pro-GMO network changed the law. The main instrument for that accomplishment was the increase in institutional power by mobilizing the agrarian lobby in the parliament and then expanding its influence to other organizations. Indeed, the powerful congressional front that systematically promotes and defends the interests of agribusiness – formalized in the Brazilian political system as the *bancada ruralista* – was crucial for tipping the balance in direction of a pro-GMO policy. It was the National Congress that approved GM soy and overturned the judicial ruling with the moratorium. When the regulatory commission could not reach a consensus to approve new products, it was the National Congress again that changed the voting rules and promoted a fast-track regulation of GMOs. Due to their increased representation in the governing coalition, politicians acting in the interests of agribusiness also dominated the executive power, where disagreements among state bodies were silenced from a top-down decision. Finally, in regards to discursive power, the National Congress was also a main locus to criminalize agrarian movements and their ties to the Lula government. They managed to taint the public legitimacy of public policies aimed at strengthening the conditions for alternative agrarian policies to flourish, such as land reform and credits for family farming. The pro-GMO network also pursued a media relations campaign in 2004, which mass media publicized; however, media coverage remained plural, advocating liberalization of GMOs together with precaution and regulation.

In the end, albeit social contestation over GMOs continued, the coordinated strategies of the pro-GMO actors' network led to the outcome of GMO dominance. After the legalization of GM soy and the approval of GM corn, these came to dominate agrarian production. It was not hegemonic, however. This confirms the importance of other dimensions of power for the material domination. There was some progress in state policies promoting alternative agrarian models based on family farming and agroecology. The issue of GMOs remained a controversial topic in the political agenda and in the public debate. Mass media covered parliamentary disputes and public disagreements among members of the regulatory commission. The domination over plant research and production had to coexist with a non-GMO market, which responded both to demands from external markets and to farmers demanding solutions to increased costs due to pest resistance among GM crops. Another sign of the absence of hegemony was that the pro-GMO network was not satisfied with the legal framework: the *bancada ruralista* constantly launched bill projects to dismantle legal provisions that guaranteed some room for a precautionary approach, for consumer rights, and, above all, for public participation in the policy of GMOs. Had it not been for a public hearing in which rice farmers and agronomists could express their views, the regulatory commission would most likely have approved GM rice. This shows that social mobilization had enough influence to demand that the other side make some concessions and amend their original scripts. Even though social movements did not win most battles, their sustained opposition certainly shaped the history of GM crops in Brazil.

Material dominance versus symbolic legitimacy from agrobiotechnology

In both countries, the dominance of GMOs in agrarian production was not a result of a legitimization process by exposing the new technology to scientific scrutiny and public controversy. Rather, it was achieved by closing the scientific and public debate whenever possible. Proponents claimed science as an ideology to legitimize biotechnology while simultaneously dismissing science as praxis to search for a provisory 'truth' by exposing results to critique. It was also not legitimized by means of a public debate in the parliament, even when this did take place, as in Brazil. Although there was a highly polarized legislative process over the *Biosafety Bill*, in which deputies represented both a precautionary position over biosafety and an advocatory position, the latter eventually prevailed, with some deputies being co-opted by Monsanto, which funded a trip for politicians to South Africa and the USA. The trip provided opportunities for the formation of networks to defend the multinational interests.

Finally, the dissemination of GM seeds in both countries cannot be understood without considering the overt reliance on illegal means. In Argentina, the lack of earnings from intellectual property rights over the seed camouflaged the technology price, which influenced the farmers' decisions on costs and benefits. The widespread use of *bolsas blancas* was crucial for the dissemination of the biotech food regime in Argentina. In Brazil, seeds were smuggled from Argentina, a strategy that created a situation of *fait accompli* that shaped the history of GMOs because it created pressure to change illegality into legality. Also, the control of seed supply often left no alternative to farmers other than to buy GM seeds, a situation that was only possible because antitrust authorities oversaw their legal duties. Also, due to the absence of feasible coexistence rules, contamination between GM and non-GM crops forced farmers to convert their fields into transgenics in order to avoid legal risks of being sued by the technology proprietor. Farmers were denied their right to decide what to produce. The more farmers adopted GMOs, the more they was advertised as a sign of their superiority.

Last but not least, the material dominance of GM seeds could at first be explained – in addition to the strategies of the bio-hegemony network – by its technological performance. Indeed, productivity increased in the first years, although recent studies question the role of genetic engineering vis-à-vis other plant breeding techniques in explaining this productivity (Marin *et al.* 2014). However, this technology performance has given many signs of receding as resistance developed and costs increased to manage it. The technology failure has not been resolved nor compensated by its developer. Farmers have few opt-out possibilities for the reasons described previously: the market domination of GM seeds and the absence of rules for coexistence. In short, the statistics showing the material dominance of GMOs in Argentina and Brazil cover all the illegal ways used to achieve those results, which remain short of public and scientific scrutiny.

Explaining different paths and outcomes

Let me turn now to the key factors that explain these different outcomes in an attempt to generalize the conditions favouring the democratic participation in shaping the model of agrarian development and the conditions for closing it. In order to do so, I will first compare the two cases in their first phase to explain why Brazilians could at first mobilize and make a difference in the path of agrobiotechnology in the country and what prevented activists in Argentina from reaching a similar outcome. Then I will make in case comparisons to understand which changes in those factors made activists lose the battle in Brazil and which changes in Argentina affected their chances of shaping the food regime and what factors kept preventing them from doing so. With the inductive approach, I can now extract from the comprehensive theoretical framework three key factors: organization and network bases, meanings, and structures of opportunities for movement action.

Why did Brazilian social movements achieve a moratorium on GMOs and create a controversy about the new technology? Why could Argentineans not shape a similar path?

The first explanation is in the organizational bases and social networks of social movements, which is related to the strategies and resources that they can mobilize. It can be divided into two elements: grass-roots bases of social movements and professionalization of NGOs. The agrarian poor are the most likely segment to form a grass-root movement against GM crops because they suffer the effects of the agrarian change more directly. In Brazil, agrarian popular movements had already established local, regional and national networks and organizations in their sustained mobilization around land reform in the late 1980s and 1990s. In this cycle of contention, the MST was innovative with its disruptive forms of action such as occupying unproductive lands and public offices of irresponsive authorities. All these previously organized actors, networks and strategies were then promptly mobilized to the fight against GMOs. In Argentina, peasants, indigenous peasants and family farmers were victims of an increase in forced evictions due to the rush for land and the effects of neoliberal reforms in agrarian policies. They responded to the advancement of agribusiness by organizing themselves, forming networks and deploying disruptive action repertoires. However, these processes developed in reaction to the strong advance of the neoliberal biotech food regime in the countryside that became visible in the expansion of GM soy fields. Therefore, the pre-existence of networks and organizations among agrarian poor that often pre-mobilized due to larger cycles of contention is a necessary factor to explain the Brazilian outcome.

Among the urban-based professional NGOs, activists in both countries were embedded in transnational advocacy networks forming the global anti-GMO coalition in areas such as environment, agroecology and genetic resources, and consumer rights. In the early 1990s, contacts between the transnational organization GRAIN and the Argentinean agroecology organization CETAAR led to a communication front, spreading the word on the risks of the new technology; however, it did not take the form of a campaign. In 1995, Greenpeace launched a global campaign, inviting countries to participate and start national campaigns in places where the technology was spreading. The Argentinean Greenpeace office decided not to start their own national campaign, although they took up actions against GMOs, since they considered that they did not have the capacity to mobilize large segments of the population.

Greenpeace Brazil saw in the topic of GMOs a possibility to connect distant forests and seas to the urban landscape, and launched a campaign that lasted for 15 years. The media-designed actions of Greenpeace, which enjoyed recognition in international press agencies, influenced the Brazilian media to cover them. The ship blockade of 1997 was followed by a judicial action against GM imports. Indeed, the use of legal mobilization has marked the Brazilian campaign against GM crops since its very beginning. The IDEC acted promptly with an injunction against the liberalization of GM soy before its official approval in 1998. In short, professionalization in the use of legal mobilization targeting national politics as well as in the design of and media actions impacting national mass media were crucial resources that affected the initial success of Brazilian activists.

The second explanation lies in the symbolic construction of critical meanings over GM seeds and the agrarian change embedded in their arrival. GMOs were a complex technology and framing it was a difficult task in itself. Movements had the challenge of combining the framing of something new with more familiar contexts. In Brazil, there was a critical mass among the rural poor that provided a background for interpreting the introduction of GM seeds as part of an agrarian model that would further jeopardize their fight for land rights. During the 1990s, the MST was Brazil's largest grass-roots movement, enjoying public support and even some romanticized coverage in national media. In Argentina, critical interpretations were formed as a reaction to the agrarian transformations. At first, small family farmers expected that by adopting the new technological package they would increase productivity and income, and thus ascend socially. Although there were many family farmers who benefited from the new model, thus becoming its supporters through their associations, many did not manage to make ends meet and had their lands confiscated. Female farmers started to critically interpret the political reasons behind their situation and this was the first step for their mobilization. For peasants and indigenous peasants, GM soybeans became the visible face of the neoliberal biotech food regime and they started to construct collective action frames in terms of land rights, environmental and health protection. However, these actors were invisible in the main media. A difference in meaning-making is thus related – although analytically independent – to the previous mobilizing structures as these provided the context in which collective action frames against the new seeds were socially constructed, as well as to the existence of a collective actor with a strong media standing and public credibility such as the MST. In other words, not only what is said but who says it makes a difference in the chances of having the message delivered.

As professional NGOs from both countries participated in the same transnational advocacy networks and had contacts to the global anti-GMO movement, they were part of a global process of critical meaning-making. The organizations that joined the GM-Free Brazil campaign in 1999 brought a wide spectrum of frames: environment, consumer rights, health, human rights and democracy. Creating a bridge between those areas resonated with an urban constituency. This did not happen in Argentina, although these frames were also part of the discursive strategies of organizations. Activists recall that there was a relatively low interest in environmental concerns in the society and less media space for covering environmental topics.[1] In short, the difference here lies in the display of strategies that paved the way for opportunities for social movements to have their views reaching the public sphere. This depends, in turn, on the structure of opportunities for movement action, to which I now turn.

The structure of opportunities for movement action was a decisive factor in explaining the success of Brazilian movements at this first stage. It included legal opportunities, access to decision-making arenas, discursive opportunities, influential allies in politics and science, and divisions within the state. Although the democratic transition meant the expansion, in both countries, of legal opportunities for social movements fighting for collective rights and public interest, only in Brazil were these used immediately to stop the adoption of GM crops. The ship blockade from Greenpeace produced allies in the judiciary system and the Public Prosecutor's Office, resulting in the first injunction against GM crops, specifically imports. But the most significant outcome from Brazilian activism came a year later, from a legal injunction that established a moratorium on GM soy. The success of such action is deeply related to the opening of institutional channels for civil society representatives in decision-making in Brazil. While Argentinean agricultural authorities approved GM soy in 1996 in a bureaucratically insulated way that was guided by administrative acts, in Brazil experiments in participatory democracy paved the way for the invitation of a consumer rights' organization (IDEC) to participate in the state commission to regulate biotechnology. The success of their injunction brought the issue to the media and many activists recall this moment as a golden era of public debate. In short, the first movement actions in Brazil found discursive opportunities in the media. In Argentina, the regulatory commission also invited a representative from civil society but chose instead an environmentalist organization characterized by its scientific expertise not for its activism in the defence of citizens' rights.

Additionally, activists in Brazil counted the Workers' Party among their influential allies in the political system. It had a foundational link to the demands of family farmers as well as an ideological commitment to the fight against the neoliberal agenda. In 1999, the PT came to the head of the state government in Rio Grande do Sul and declared the intention to make it a GM-free zone, challenging the national government. This state would later become the stage of the first World Social Forum in 2001, where important transnational protests against GM crops took place.[2] The scientific community also was an influential ally to social movements because they defended a precautionary approach. Finally, in Brazil, networks of activists were represented in state bureaucracy and created a division within the state that challenged the Ministry of Agriculture's promotion of biotechnology. State bodies competent for health, environment, consumer rights and family farming served as internal allies in the dispute over state policy. In this early stage of mobilization in which alliances were necessary to construct an issue in the political agenda, Argentinean activists recall their difficulty in recruiting experts and allies in the political system. The Argentinean state had a unitary position in favour of GMOs and did not provide openings to debate the issue.

The three factors combined and their timing resulted in a path of mobilization that shaped contentious politics over GM crops in Brazil. It was crucial that Brazilian activists acted early to throw their opponents off guard. The blocking of a shipment and the use of an injunction to suspend the approval of GM seeds proved disruptive to the pro-GMO coalition's original plans. It was not in their script. As dominant as the advocates of GM seeds had been in the USA and Argentina, they did not expect to be met by contentious Brazilians. The early risers in Brazil thus created opportunities for the opposition of this new technology, inspiring other movements to start a national campaign. In 1999, the GM-Free Brazil campaign was launched, organizing collective actions beyond single events and representing a broad coalition of NGOs and movements when making claims to authorities or contacting possible allies. They established a mobilization front, a communication front and a judicial front. The coalition diversified resources and strategies, enabling the movement's campaign to use some leverage in venues where their powerful enemies were well represented, that is to say, in the executive power, the parliament, the judicial system and the commission of experts.

By contrast, the pioneer role of Argentina in adopting GMOs made it even harder for movements to preventively act; rather, they reacted to the consequences of the already widespread dominance of GMOs, often building mobilizing structures and critical meanings to GMOs as processes of reaction. Their contentious meanings and strategies rarely found allies and opportunities to reach the wider public and a political agenda. A path of mobilization against GMOs was even more unlikely to unfold because Argentinean activists had a hard time in building long-lasting coalitions. While different organizations cooperated in some actions in Argentina, this was rather the exception than the rule. There is an old and lasting network of activists in agroecology from which important organizations had emerged. While showing solidarity, they did not join collective efforts in a campaign against GMOs.

Why did contentious Brazilians lose the battle?

A way to answer this question is to look at what has changed in the three main explanatory factors. In regards to the organizational bases and social networks of social movements, there was a new factor: the brokerage between agrarian movements and NGOs. Aware of the political potential of contemporary peasant movements, the organizations of GM-Free Brazil targeted framing strategies to bring them into the campaign. Their adherence brought grass-roots mobilization potential. For peasant movements, joining a campaign against GM crops meant an important alliance with experts who were not mobilized to support their land struggles previously. This was particularly important when GM corn reached the regulatory agenda and there was a joint mobilization against its approval, including a wide repertoire of actions, such as lobbying and participation in the expert commission by mobilizing expertise, protests and legal action.

This alliance was reflected in the construction of contentious meanings: agrarian alternatives received more prominence in the discourse of GM-Free Brazil, which incorporated the master frame from Via Campesina. However, there was a main change in relation to the previous phase: the reputation of the MST was under heavy attack. As Lula assumed, he received the MST to listen to the movements' demands, signalling the openness of his government to them. Authorities and elites organized a strong reaction to threaten the MST since they feared their movement's power and the influence of the movements from Via Campesina in Lula's government. Agrarian elites used their influence in state institutions to find legal means to control peasant movements, thereby increasing the costs of their mobilization, a mechanism that social movement scholars have called 'channelling' (Tarrow 2011). More than that, it entailed a process of criminalization by state institutions and the media that attacked peasants' public legitimacy, particularly the MST. These reactions from authorities and elites contributed to a process of demobilization.

Another hard blow suffered by agrarian movements in their fight against GMOs was the gradual adoption of GM soy among their bases. As in Argentina, family farmers in Brazil also wanted to try the promises of biotechnology and even smuggled GM seeds. But this type of reaction coexisted with a critique on the narrative of technological modernization and with peasants' movements against it. When peasants started to follow the path of technological modernization by planting GM soy too, it was a blow for their leaders. It was much harder to campaign against GM crops: the fact that some MST camps were also using GM seeds affected the movements' legitimacy for contesting it; indeed, the media widely explored the issue. Peasants' movements had to wait until the technology showed its limits and their bases started to realize the higher costs and negative effects of GM soy.

Nevertheless, agrarian movements kept following the issue, together with the NGOs. As the *Biosafety Bill* passed, activists started to monitor the expert commission. The shift of attention to the regulatory arena required a stronger description in legal and scientific terms. In particular, claims for legality, transparency, accountability and democracy were recurrent, showing how the application of the law had to be disputed with a combination of direct action and legal mobilization. When corn reached the agenda of the CTNBio, peasant movements became alarmed. The eminence of GM corn provided a clear threat to farmers' rights over seeds, with the fear of having to pay fines for having their fields contaminated by cross-fertilization and find the patented gene among their seed mix. Contesting GM corn found resonance and was influential due to its empirical credibility and experiential commensurability. It also mobilized emotions due to the family traditions of corn production. Food sovereignty became a prominent master frame in this moment. Finally, GM corn mobilized gendered identities of mothers to the issue, with frame alignment activities highlighting the health risks to pregnant women consuming GM corn. The social mobilization against GMOs in Brazil thus constantly evolved with the development of contentious meanings about GM crops and their consequences. However, their discursive constructions as well as their media-designed strategies and legal action had less and less public resonance. This is related to the next factor.

The structure of opportunities for movement action had been transformed by a key change in the type of alliances and by the mending of the divisions in the state. Concerning the former, there was a transformation from institutional to individual alliances. What had previously been institutional support since scientific organizations, a political party and some state bodies allied with social movements turned into an official position either favourable to GMOs or silent, discharging members for their 'personal' views. The clearest example of this change was PT. There was a rift in the party, and GM-Free Brazil lost an ally that became increasingly more influential because the political party would dominate national politics in the years to come. Therefore, the GM-Free Brazil campaign now had the support of individual politicians, individual scientists and individual state bureaucrats but lost the influential alliances with organizations.

The second change was in the governmental position regarding GMOs. The GM-Free Brazil campaign soon realized that Lula and the PT, once in office, lost their commitment to the cause that they expressed when in the opposition. As Lula signed already in his first year in office a provisional measure liberalizing the commercialization of illegal GM soy, activists reacted with legal mobilization; nevertheless, it had no effect in contrast to the immediate effects of Lula's legitimation of the politics of *fait accompli*. The pro-GMO coalition explored the government's ambiguities in not taking a clear policy decision and in order to ensure control over the GM policy placed the issue on the legislative agenda. Activists disputed every word of the text of the *Biosafety Bill*, aiming to influence a decision-making process guided by the precautionary principle and by democratic rules. Notwithstanding their efforts and some of the concessions they achieved, the 2005 *Biosafety Bill* formalized a policy favourable to agrobiotechnology and to the official transformation of Brazil into a GM crops producer, reversing social movements' outcome of the last phase. As the main arena to dispute the GM policy shifted to the commission of experts, where there was a polarization between two risk cultures, activists sided with allies who defended the precautionary principle. When the CTNBio approved GM corn, activists took a legal action and motivated the support of state agencies on health and environment, which used their legal prerogatives to appeal the decision. However, the CNBS, having been forced to deliberate over the appeal, decided to give the final word in the decision on GMOs to the CTNBio. This was a top-down political decision and meant the closure of state policy in favour of GM crops, silencing the positions of other state bodies and favouring an expert culture that promoted GM crops and neglected scientific evidence of risks.

More than any other factor, it was the progressive closure in the structure of opportunities for movements' demands that explains the change in the outcome, namely the conversion of Brazil into a top world producer of GM crops. It also contributed to a path of demobilization in Brazil. After the approval of GM soy and GM corn in Brazil, social movements realized that there were no openings in the political landscape for opposing GMOs and that the discursive opportunities in the media also diminished. This context helped to build the perception among social movements that the probabilities of achieving positive outcomes in the struggle against biotechnology were low. After almost ten years of fighting, this perception of loss helped to build a situation of exhaustion due to the strategy of politics of *fait accompli* used by the bio-hegemony project. They felt that no matter how much they fought there was no chance of achieving their goals because pro-GMO decisions were already taken. This perception contributed to a process of demobilization of social movements and some organizations, like Via Campesina and Greenpeace, ended their own campaign on GMOs, although they did not leave the coalition of GM-Free Brazil.

Timing again was an important factor in the trajectory of Brazilian contentious politics over GMOs. The initial success of social movements' campaigns is also an important part in the explanation for their later reversal when GM crops were approved. This is because the pro-GMO coalition learned from this experience and employed more resources to achieve their goals. However, by having acted early, before the policy was closed, their activism shaped that policy nevertheless: this is shown in the provisions for transparency, public participation and the precautionary principle in the *Biosafety Bill*. Activists' pressure on the CTNBio was what made it issue regulations and respect some criteria of procedural democracy. Although the novelty and dramatic character of the issue receded, thus diminishing media attention, activists managed to have their views covered anytime a new event occurred, such as the audience on GM rice in 2009. Moreover, the past years of mobilization had been fundamental in consolidating a network of activists and organizations that shared the goal of a GM-free Brazil, thereby building relationships of trust. Activists who remained active in the campaign could mobilize their partners when support was needed. When the biotech food regime materially dominated the Brazilian fields, activists remained alert concerning its consequences and promptly joined efforts in a campaign against pesticides.

Why did Argentineans, although mobilizing, not challenge bio-hegemony?

The comparison between the two phases of the trajectory of GM crops in Argentina shows substantial changes in the first and the second explanatory factors, underlying an increasing mobilization, but no significant change in the third one, which relates to the structural opportunities for that mobilization to strive and influence society and politics. Organizational bases and social networks of social movements were mobilized around three issues related to the agrarian model. First, this period marked the consolidation of the national peasant movement in Argentina, which bridged the identities of peasants and indigenous people, and claimed a different way to relate to land and nature. Proposing a productive model of agroecology and peasant farming, they differentiated themselves from family farmers that pursue small-scale agribusiness. They inscribed themselves in a non-capitalist farming model, relying strongly on a class identity.

Nevertheless, their production faced legal and administrative hurdles due to a lack of adequate policies. Above all, they were confronted with the immediate need to guarantee their land rights. The peasant movement has first concentrated on a defensive strategy of disruptive actions to stop forced evictions, and started to diversify topics and action repertoires while they engaged in lobbying for a bill project to stop forced evictions, and with agendas such as the seed law, for which they organized protest events.

A second node of mobilizing structures also took part among the rural poor and those living on the margins of the model, with grass-roots initiatives from groups of neighbours affected by pesticide spraying in regards to their health and environment. Various locally organized neighbourhoods established a national advocacy network of sprayed peoples. They were innovative in how they mobilized counter-expertise; they organized protest events and took legal actions targeted at the local and provincial levels due to the decentralized structure of the Argentinean state in regards to competence areas of health and environment. A third issue was forests. It gained momentum around a brokerage between grass-roots mobilization among neighbours in places suffering from local issues and a professional Greenpeace media campaign to stop deforestation. They combined the use of disruptive actions and political lobbying, which culminated in a national forest law in 2007.

This growth in the organizational bases and networks of social movements went hand in hand with a stronger meaning-making work of the causes and solutions for their problems while they created a critical interpretation of the agrarian model in Argentina. The master frame of food sovereignty spread from peasant movements to urban organizations, helping them to frame the negative consequences for labour, health and environment of the model oriented to export commodities; particularly dramatic was the contrast between food and commodity. Among the sprayed peoples, the construction of collective action frames that linked their health problems to the soy model was crucial for their mobilization because it meant changing the attribution of individual suffering from fate or individual responsibility to the diagnosis of a social problem caused by the agrarian model. Their framing also attributed responsibility to the state owing to its lack of enforcement of environmental laws. The Mothers of Ituzaingó developed a key strategy in the fight against the agrarian model: they mobilized counter-expertise by building their own epidemiological map. For that, they relied on resources that were not readily available for others: through their conviviality in the neighbourhood, they had access to peoples' homes and were trusted in sharing their family suffering. The Mothers of Ituzaingó also brought emotions to the movement and mobilized gendered identity politics. The strong symbolism of women and politics in Argentina, in large part due to the movement Mothers of the Plaza de Mayo guaranteed a high resonance for their claims. Their moral standing helped to build solidarity among many activists.

Such mounting mobilization, diversification of strategies and ongoing symbolic work have not resulted in the transformation of movements' demands into a wider public debate or the construction of a social problem. There continue to be significant obstacles in the structure of opportunities for their action despite some changes. Key changes took place in two components: influential allies and legal opportunities. Circumstantial openings occurred in the state and the media, to which the pro-GMO network reacted and in turn shut them down. There was an apparent division among elites but it revealed rather that the political landscape was closed to the movements' demands.

The peasant movement found new allies among agronomists with the opening of the Free Chair on Food Sovereignty in the University of Buenos Aires, which meant a small fracture in the mainstream agribusiness model at universities, and also increased their access to the Ministry of Agriculture with an ally in the Under-Secretary of Rural Development and Family Farming, also a small entry in an otherwise bio-hegemonic institution. The pesticide issue brought the support of professional groups such as physicians and lawyers to the debate over the consequences of the agrarian model. These new influential allies to social movements would strengthen the action repertoires of legal mobilization and counter-expertise. As an influential scientist, Carrasco gave an interview in the newspaper about the negative effects of glyphosate and for the first time there was a fissure among the national executive power as one ministry openly assumed a critical position against glyphosate. It was short-lived and soon dismissed, but it did leave traces in the mediated debate. Even if most media groups reacted with censorship, it nevertheless opened discursive opportunities to relate glyphosate to negative health effects. However, Carrasco was met with threats and coercive responses from political authorities, which attempted to restrict his influence, and with attacks from agrarian technical chambers and mass media groups, which tried to attack his scientific reputation. Such episodes reveal the constraints faced by scientists and public officials with dissident views in Argentina regarding the biotech food regime.

Legal opportunities opened to Argentinean movements as the court ruling of Ituzaingó Anexo associated, for the first time, agrarian activities with an environmental crime, having condemned individual farmers and issued recommendations for the executive power. Their legal victory showed others that if the political system was closed to their demands there were other institutional arenas to explore. Having reached the court, they motivated allies in the judicial system, among them lawyers, doctors and scientists, and a few politicians and public officials. They received support from the Mothers of Plaza de Mayo, among the most influential human rights organizations in Argentina, also showing the prominence of the historical frame of women and human rights in Argentinean politics.

The new alliances and the court ruling were not enough to bring activists' demands to the national political agenda or public debate. They found no points of entry in the state, which remained closed in the defence of the model. In 2008, Argentina experienced a major division among its elites known as 'the rural battle'. The conflict between the elite in power and the representatives of agrarian producers regarding export taxes brought the agrobiotechnological model to the top of the public agenda. Activists attempted to transform this episode into a window of opportunity for discussing the model. However, in the end, the conflict revealed a political alignment concerning state promotion of agribusiness: what was under dispute was the distribution of its rents, with state dependence on soy exports openly acknowledged, whereas the model was not called into question. The division among elites exposed to activists that in fact there was an underlying consensus among the political and economic elites that GM soy was good for them and for Argentina. In other words, it revealed that bio-hegemony reigned in the country.

In sum, the changes in organizational bases and meaning construction were not sufficient to build a public controversy and debate over the agrarian model in Argentina due to the fact that the structure of opportunities for their action remained closed. However, these changes set in motion a path of increasing social mobilization against consequences of the widespread adoption of GMOs. Such mobilization has already proved to be an important mechanism in creating opportunities for movements in Argentina: it drew some influential allies as well as contributed to the diffusion of frames and the formation of collective identities and to some opening in the judiciary power. Mounting mobilization might trigger more opportunities for social movements. Some activists argued that a new cycle of contention was in the making. They saw that prospects for mobilization were increasing, in particular in sectors of directly affected groups, such as mining and spraying activities. All agreed about the leading role of the Mothers of Ituzaingó as early risers who, with their mobilization, not only created opportunities for their fight but also promoted the diffusion of opportunities by showing to others that it was possible to do so, even for a group of mothers from a poor neighbourhood. Although it was considered impossible to contend such agricultural practices, their maintained campaign has slowly revealed points of weaknesses in the agrarian model that could jeopardize its legitimacy in the Argentinean society. Not everyone shared this rather optimistic prognosis. Carrasco, for instance, noted that there was no mobilization in the urban settings and the national capital against the model but rather a resistance in the affected territories. He also highlighted the strong obstacles to target agrarian activities, compared to mining, as farmers have been historically positively framed in the Argentinean society. This once again underlines the discursive power of the agrarian model in Argentina.

In short, the pioneer timing of bio-hegemony actors' network helped them to guarantee enough material, institutional and discursive powers to obstruct the structure of opportunities for challengers. When activists found allies or media opportunities, they were met with vehement and concerted attacks from the bio-hegemonic establishment. Even the judicial ruling of 2012, with multiple recommendations to the national authorities, fell flat in the emptied political landscape, not receptive to critiques of the agrarian model.

How did a new context in the political economy affect movements' opportunities in Argentina and Brazil?

Context conditions help to explain the key role the structure of opportunities played in defining outcomes in the second phase. In both countries, this period was characterized by the closing and consolidation of an unambiguous state policy promoting GM crops as the main obstacle for social movements' efforts to make this issue the object of public debate and deliberation. This reinforced consensus among top decision-makers and the political and economic elites in favour of GMOs is deeply related to a change in the context of the political economy of these countries. Changes in two context conditions are particularly relevant for interpreting the new structure of opportunities and constraints for social mobilization against GM crops. The first is the increased reliance on the production of commodities for export, coinciding with an increase in the prices of global commodities. This augmented the economic and political dominance of agribusiness. The second point is that both governments have invested an increasing amount of state revenues – which came in large part from commodities – to expand the coverage of social policies and achieved results in poverty reduction, which had demobilizing effects.

As new governments were elected in 2003, it was not clear for any political actor how effectively progressive they would be and how they would translate their programmes, discourses and structure of political alliances into actual public policies. The conservative forces, however, did not expect to have to create political space in the definition of policies that were most important to them. This included the agrarian policy. In Brazil, the PT government did not negotiate, however, in its main programmatic statement: to eradicate poverty. As long as it could pursue this main project, others were open to negotiation. In Argentina, the immediate priority was to restore the country from the effects of the economic and political crisis, among them the high levels of unemployment and poverty. In both cases, the increase in state revenues due to commodity exports was the easiest and most direct way to realize their goals. This coincided with a period of ascending prices of commodities – among them, soybeans – in the financial markets, making the option of promoting commodity production even more plausible for the new elected governments in need of revenues to fulfil their priorities.

Social movements contesting GM crops faced an ambiguous context to take their actions. In Argentina, other movement sectors had increased participation in government, but this was not the case for the movements studied. In Brazil, there were institutional openings for agrarian social movements with an increase in channels for dialogue, nomination of some movement leaders to state bodies and increase of public funding for social movement organizations to undertake social programmes. But dialogue was not translated into effective political actions that would advance social movements' agenda for an alternative model of agrarian development. The government maintained some space at the margins for advancing the agenda of agrarian popular movements as long as it did not challenge the core. Even less political determination was found among the new incumbent to deter the expansion of agribusiness. In both countries, opposition to it was controlled or even suppressed. Agrarian elites and also multinational corporations made use of outright coercion, threats and went as far as to kill movement leaders. Their will to control dissent had the backing of state authorities while their human rights violations and crimes were left in impunity. The record of violence in the countryside is another indication of the unequal application of the law. In Brazil, the MST faced a renewed wave of criminalization, which posed obstacles to their mobilization.

The second context condition that influenced mobilization was the effects of social policies and income growth among the grass-roots basis of most agrarian movements. This had demobilizing effects according to their leaders. In Brazil, the PT government ensured that social policies of cash transfer reached the countryside and improved access to basic public services such as electricity. However, the more significant impact came from the real increase of the minimum wage because the rural poor have the right to public pensions indexed to the minimum wage. In Argentina, the government contracted social organizations to distribute social programmes, reaching many in the poorest regions of the country, among which were the bases of agrarian movements. The lives of the agrarian poor improved since they had access to a minimum wage and some public services. However, their capacity to maintain their livelihoods through their own work was severely affected not only by the expansion of agribusiness – which brought with it displacements, chemical contamination, reduction of biodiversity and violence – but, above all, by the absence of proper public policies to allow them the means of pursuing their agrarian activities according to their way of relating to land and nature. In Brazil, the policy for land reform and land entitlements for ethnic communities had lost momentum in their implementation; in Argentina, it had barely started. In both countries, leaders from peasant movements claimed that they would prefer for the government to pursue public policies favouring land reform and peasant farming instead of having to rely on programmes of conditional cash transfer.

As elections approached, however, when movement leaders had to inform their political support to the bases, they instructed them to vote for the incumbent government due to fear of the opponents reversing the improvements of their living conditions. Therefore, the movements of the agrarian poor remained an important ally of *lulismo* and *kirchnerismo*, which revealed how they were afraid of losing even more with an oppositional government more aligned with agribusiness. In Argentina, peasant movements thus fell into a complex blame game. This is also explained by more stable aspects that influence contentious politics in that country, a mix of a centralized state in respect to some policy attributions with a decentralization of policies that speak to the use of natural resources, including their health and environmental impact. As a result, there is a multitude of targets on various scales. In many cases, movements prefer to blame local and provincial authorities rather than the national government. Multinationals like Monsanto became favourite targets. Although greedy corporations often provide a powerful frame (Gamson and Modigliani 1989), this can also be an obstacle to redress their situation, since attributing responsibility for corrective action cannot evade the question of the role of the executive in fostering the agrarian model.

In both countries, there have been continuous disputes inside social movement groups on how to make claims to authorities as well as on the use of disruptive action repertoires. In hindsight, social movements made a vital error in having misread the structure of political opportunities. In Argentina, they failed to see a change in the new government in 2003 that could have been better explored in furthering their goals. In Brazil, they had difficulties in understanding and taking part in the 'government in dispute', ignoring that the government itself faced dilemmas and that more social struggle could increase pressure for a more progressive agenda. However, it is hard to assess whether they would have achieved more given the context of increasing state reliance on revenues from commodities – closing the extractive and agrarian policies to any criticism – and the difficulty of finding influential alliances in this context, as described in the previous sections.

In addition to such material state dependence on the extraction of commodities for export, activists in both countries identified that the government, although with a progressive social agenda, defended a developmentalist conception that was susceptible to the idea of progress and economic growth, without considering the ecological and social impacts of its policies. This goes well beyond the topic studied here – expansion of the biotech food regime – and includes all other extractive industries that have been promoted during *kirchnerismo* and the PT government, such as mining, energy exploration and enormous infrastructure projects. The electric power plant of Belo Monte in the Brazilian Amazon is a symbol of the neglect of indigenous rights and the rights of traditional communities whose livelihoods depend on that environment, as well as of unforeseen environmental impacts. In part, the social bases of these governments, which have historical ties to and wide support among urban workers and organized unions, also share this conception of development in which more and more segments of society are included in the labour and consumption market. For them, the places in which commodity extraction takes place are distant abstractions; for many, peasants as well as indigenous and traditional peoples should have the opportunity to also be included in the labour and consumption markets. Nevertheless, one cannot say that such conception is a product of or related to the successes of the strategies from the bio-hegemonic network. Rather, it appears to be a remnant of the old left conception of growth and full employment as a means to include the poorest segments in the labour market and from there on advance towards other demands. It is a powerful discourse, in particular in a context in which exclusion from the formal market and from consumption has been the rule. Indeed, the comparison with the previous phase shows that neoliberal adjustments provided a much clearer target for civil society mobilization and coalition building across urban centres and the countryside, whereas the recent context shows a state with a much more ambivalent face for subaltern sectors and an increased gap between urban and rural grievances.

*How did the integration of GM crops in the production node of global
commodity chains affect social movements' opportunities?*

The increasing reliance on commodity exports by the new governments in Argentina and Brazil raised the question of how the integration of their agrarian model in global commodity chains affects movements' campaigns against GM crops. Biotechnological corporations decided to focus their GM seeds on crops that are embedded in global commodity chains as inputs for animal feed, industrial foods, biofuels and industrial uses. The countries studied have specialized in the production of commodities that are flex crops for global chains, in which oligopolies of transnational corporations control important nodes (inputs, commercialization, processing and retailing). The majority of GM crops produced in Argentina and Brazil are destined for export markets. In other words, there is no GM food produced for direct human consumption on local markets yet. This position in the global commodity chain has various implications for activism.

In both countries, activists targeted national consumers in an early phase of the arrival of GM crops. However, there was an abyss between what happened in the countryside and in the urban centre since the agrarian production was directly integrated into external markets and hardly visible in urban food consumption. Brazilian campaigns were more successful in making GM ingredients visible to urban consumers through the use of tests on GM ingredients in processed foods, which are products for direct consumption. In addition to consumers, activists in both countries targeted capital by making demands on national and transnational food industries. In Argentina, this resulted in a local ordinance requiring labelling, which the federal state overturned. Brazil approved a labelling law but it is debatable whether its low enforcement has had any effect on consumer behaviour. More common is that only consumers who are already part of a food culture sharing ecological values or concerns about health will actively choose GM-free products, which in both countries would be restricted to small segments of more affluent consumers. It did, however, meant a legal opportunity for activists to monitor the industry, thus placing an onus on the food industry because it had to make its supply decisions public while making statements and responding to the lists periodically published by activists of GM ingredients in processed foods.

Both in Argentina and Brazil, risk discourse resonated more among the poor rural constituencies, where production takes place, than among the urban consumers. The social question does not come first as in a linear model that would forward that risk concerns would have to wait for the resolution of the social question (Beck 1986, 2007). Rather, the cases studied show that concerns about risk deeply intersect social issues: in line with findings on environmental justice (Martínez Alier 2003; Porto and Milanez 2009; Tesh 2000), the poor are those who suffer the effects of GM crops the most in relation to physical health and environment. More than targets of frame alignment activities, they have themselves started to make sense of the very tangible damage experienced. Risk becomes material. Popular epistemology establishes correlations between the expansion of GM crops and their sufferings; they do not depend on scientists in distant laboratories to calculate probabilities based on a few rats. Activists did not have success in warning the urban middle classes about abstract risks without scientific proof, while the present damage in rural communities motivated them to dispute dominant risk assessments. The time dimension, therefore, was crucial and favoured the proponents of the technology.

Political economy and phenomenology are brought together because sprayed people and peasants directly experience the violence brought when global market dynamics ally with national development projects. People living close to GM crops resorted to subaltern strategies of place-based identity construction in defence of their 'territories' threatened by dominant interests as lands were converted into cropped areas due to an ever-expanding global demand. Peasants contested a nationalist framing of the land issue, bringing the argument that the flag from the capital was not relevant; as long as it was capitalist agribusiness, it threatened peasant farming. The more agribusiness advanced its control over land, the stronger peasants and indigenous peasants shaped their identity by way of their praxis in connection to the land on which they were living and the way they use it. This means that the most potential target for challenging GMOs was not a node in itself in the global commodity chain that could affect the decisions in the precedent or subsequent nodes, as Schurman and Munro (2009) proposed by considering the global commodity chain as a structure of opportunities. Rather, it was the rural poor living at the margins of the production node, being expelled through the advancement of agrarian production destined for these global markets. At the same time, the adaptation of the rural poor to being sprayed by pesticides or their accommodation (Lapegna 2013b, 2014) by planting GM crops themselves have also shown to contribute to the construction of the bio-hegemony.

This brings to the fore the spatial inequalities in the distribution of the benefits and burdens of the large-scale production of flex crops: while urban affluent consumers have access to animal products, industrial food, flex cars and airplane travel using biofuels, they are distant from the places where the socio-environmental consequences are felt by the rural poor. The latter make claims for social and environmental justice, and hope to receive support from the former if they are persuaded, with scientific proof, that their consumption patterns might create health risks or, in the case of political consumers, can have ethical consequences. Given that the social, environmental and health consequences of the global commodity chain are concentrated spatially in the locus of agrarian production, the question regarding responsibilities becomes of central significance. The court case of the Mothers of Ituzaingó revealed more than only the problems for Argentina in regulating the local health and environmental effects of GM soy crops. It has shown that these are global regulatory challenges, with the burden asymmetrically distributed in the global commodity chain, but with the common denominator that most national governments have proven weak vis-à-vis the global interests of chemical, seed and food industries. This also reflects how transnational corporations can capture international institutions, as can be seen with the safety standards for pesticides established by the *Codex Alimentarius*, which acts as a powerful weapon to deny damage caused by chemical-intensive agrarian activities. The Argentinean scientist Carrasco moved between scales, travelling to sprayed communities and to Germany and Brussels, aiming at influencing the German rapporteur to the EU to take into account the local effects of glyphosate on producer countries when reassessing the toxicity of the pesticide. Although prohibiting the large-scale use of glyphosate, Europe, a big importer of GMOs produced with glyphosate elsewhere, is not concerned with its local effects on producer countries.

One could conclude that, structurally, Argentina and Brazil are in very similar positions in terms of the global commodity chain. However, actors read such structural positions in the global market in combination with the national political opportunity structure. Activism against GM crops, even when part of a global anti-GMO movement, is domestically rooted (Tarrow 2005) and when it comes to influencing state policies it takes place according to national processes of agenda formation. It relates to other issues on the agenda and other parts in the conflict, which are dynamic processes.

At first, both in Brazil and in Argentina, activists targeted national constituencies and the national government. As in other cases of contentious politics in a transnational context, activists adopted strategies of 'domestication', refocusing political attention to the national state and its sovereignty of drawing boundaries for the flows of global capitals, goods and services (Tarrow 2011). This has remained the main locus of blaming and activity in the case of the Brazilian campaign. In Argentina, the perception of a closed political opportunity structure has made it more reasonable to blame Monsanto and to appeal to European consumers and authorities. This pattern coincides with what other scholars have noted, namely that 'externalization' strategies follow from the failure of domestication ones (Keck and Sikkink 1998; Tarrow 2011). Of course, these can be made in an articulated fashion when, for instance, national governments are in alliance with subnational and local authorities as well as in alliance with transnational corporations. Argentinean activists also scaled down their actions by targeting subnational states, blaming the governor or demanding corrective action at the local level.

Brazilian activists were convinced that there were political opportunities for their agency, in general, and that the main scale of their struggle is national. They have consistently targeted the national government even though they also shifted scales between subnational, national and global levels. By contrast, the consciousness in Argentina of the fact that GM soy production is embedded in a global commodity chain, together with the widespread acknowledgement that soy exports have been central to the country's economic recovery from a deep financial crisis, provide the basis for their perception of the restricted political opportunities for action. The vulnerability to (global) market forces during the crisis of 2001 was a significant experience with many consequences. Indeed, some activists frame it as a (global) market-dependent fight, not a political fight. However, all interviewed activists call attention to the fact that the country's insertion into the global economy was a political decision based on a developmental model. In sum, Argentinean activists perceived more global constraints to their fight whereas Brazilians perceived more national opportunities to theirs.

An explanatory model for further research

The comparison between cases over time has shown that early social mobilization – with mobilizing structures and contentious meanings – is a necessary condition to participate in the shaping of the food regime; the most determinant condition, however, lies in the structure of opportunities. This finding was reached through many steps. By answering the first question, it became clear that all three conditions were necessary to explain successful mobilization in Brazil, which were more or less absent in Argentina. Addressing the second question of why the main policy outcomes achieved by social movement campaigns were reversed, it became clear that there were few significant changes in the first two explanatory factors and main transformations in the third explanation. This hints at the hypotheses that although mobilizing structures and meanings are necessary conditions to successful movement campaigns, they are not sufficient; for that, it is necessary that they find opportunities for their demands to influence public debate and policy. This was further confirmed in the answer to the third research question. In Argentina the emergence of mobilization resulted from an improvement in organizational bases and networks of social movements together with contentious meanings, but it found no opportunities to influence the outcome. In short, mobilizing structures and contentious meanings are necessary but not sufficient conditions for social mobilization to influence the trajectory of a food regime in a given country, as the latter depends on a third condition: a favourable structure of political opportunities.

This last factor is more likely to be influenced by movement action when the policy is at an early stage and its trajectory is still open. The prompt action of early risers can then trigger opportunities in the media and in the judiciary, relying on influential allies. Therefore, timing is an additional factor in explaining the chances of movements in shaping the trajectory of GM policy. Conversely, the actors' network sustaining the bio-hegemony project also benefited from early timing. However, as the pro-GMO actors have commonly more resources and access to influence public policies ('business as usual'), they are not dependent on creating opportunities; actually, their strategies to build material, institutional and discursive powers are an explanation of the closing of the political opportunity structure for movements. Finally, the latter is also influenced by the political economy, which can offer more chances for or more constraints on mobilization.

The main paths described include processes of mobilization and demobilization. Contributing to the former is the formation of a social movement campaign against GM crops and of a coalition among different organizations in order to coordinate actions beyond single events. The brokerage between struggles in the countryside and in urban settings has also been shown to be an important mechanism for a successful mobilization. The mechanisms that contribute to a path of demobilization are the adoption of GM crops by the rural poor, criminalization of social movements and violence against activists, the feeling of exhaustion due to the perception of not having opportunities to have their demands heard and met, and policies for poverty reduction.

As far as the effects of global commodity chains in social mobilization are concerned, activists in producer nodes will face harder challenges for targeting production decisions because these are important sources of private and state revenues. The agrarian poor, however, will suffer more from the global socio-environmental burden of global commodity chains, leading to their social mobilization if mobilizing structures and meanings are given. If not, the accumulation of grievances can be an opportunity to develop those factors, showing again the importance of timing as this type of mobilization will be a reaction with a time lag regarding consequences of past decisions, and thus have fewer chances to reach the political system and affect policy change. Finally, the perception that the locus of decision-making lies in national politics is enabling for social movements fighting GMOs, while the perception that there is no national autonomy for decision-making, as this is mainly determined by global market actors and forces, is demobilizing.

Notes

1 Thus, it cannot be said that there was a lack of discursive strategies on the part of Argentinean activists. On the contrary, they coined symbolically rich expressions such as 'agriculture without farmers', 'record harvest, record hunger'. The difference is that GM crops did not resonate as a symbol for those meanings in Argentina. This led Greenpeace Argentina, some years later, to completely transform the framing of the problem. The structural transformations in agriculture were subsumed under the campaign for the forests, that is to say, a consensual symbol for environmental values.
2 At the same time, it was a state where two ideologies over the model of agrarian development had clashed since the 1970s: techno-optimists and eco-socials. Thus it was a state that experienced a strong battle between the pro- and the anti-GMO coalitions, and where the former was represented by enthusiastic farmers who smuggled illegal seeds from Argentina.

Chapter 7
Conclusions

This book has narrated more than two decades of social disputes over genetically modified organisms in Argentina and Brazil to develop the argument that the transformation of Argentina and Brazil into top world producers of GM crops cannot be explained by the technological superiority of biotechnology. Rather, the trajectories behind the dissemination of GM crops in these countries were stories of political struggles over agrarian development in which social movements and the rural poor, while contesting the advancement of a biotech agrarian model, were silenced, ignored or demobilized by a network of actors in favour of GMOs. These pro-GMO actors secured the conditions for domination through illegal means by avoiding public debate and democratic participation as well as by suppressing contestation where necessary.

The playing field, that is to say, the political economy of these countries, influenced these struggles to the detriment of social movements. At first, policies of neoliberal structural adjustments severely hit the countryside while demands for protection against the 'free market forces' found no resonance in that context, particularly in Argentina. In the first decade of the century, commodity production became more and more important and increased in the share of state revenues while at the same time grain was placed among the top products of the export mix. This change is associated with a stronger state that promotes economic growth and uses commodity exports as 'an opportunity' to reduce poverty and include citizens in the consumption market. Such political economy of commodity promotion and poverty reduction provided powerful legitimacy to the development model. At the same time, the extension of agribusiness and GM crops brought with it socio-environmental damage and the diminishing of rights. However, the political choice for a development model based on growth and inclusion of citizens as consumers remains blind to its own socio-environmental consequences.

The argument of the book reveals the wider relevance of studying political struggles over GMOs for at least two reasons. Biotechnology has been portrayed as the solution to world food security – a conviction that has been supported even more after the world food crisis in 2008. Food security is a topic of dispute in the global political agenda because many movements of the agrarian poor have been radically challenging the solutions that are reliant on more technology and capital-intensive agriculture, and instead defending smaller-scale production and better conditions for access. Therefore, unravelling the complex history of the domination of GMOs in two countries that are propagated as breadbaskets of the world is informative for the wider global debate on agrarian futures and food security. The study has contributed to the debate on the paradox of increased food production with continued hunger among the world population (Magdoff *et al.* 2000). In Argentina, activists reported that 2002 was the year of record harvest as well as record hunger. Indeed, if one is to relate the production of agrarian commodities to food security, the correlation is not necessarily positive. According to the Millennium Development Goals, Argentina showed no progress from 1990 to 2010, as 5 per cent of the Argentinean population remained undernourished while the country experienced the so-called 'transgenic revolution'. In Brazil, the improvement in this indicator from 11 per cent to 6 per cent between 1991 and 2007 (Center for Global Development 2014) is thus better explained by public policies to promote access to food, such as the programme Fome Zero, rather than by the increase of commodity production. Promoting GMOs in commodities for export does not translate into feeding the world, let alone the country's own population. This raises a major objection to the argument used by social scientists who defend GMOs as a solution to food security (Herring 2007; Paarlberg 2000).

A second reason is that the issue of GMOs is also illustrative of contemporary challenges for social mobilization and demands for rights to counter systemic imperatives of global capitalism and political interests. In this concluding chapter, I will first elaborate upon the conditions that allowed social mobilization to shape the trajectories of agrarian change brought with GM crops as well as the conditions that prevented it from doing so in the cases studied. I aim at contributing to the theoretical debate on social sciences on social disputes over GMOs. Based on these considerations, I will reflect on the relevance of the study beyond the issue of GMOs, pointing to new research questions and agendas.

On the conditions for social mobilization against GM crops

Social mobilization challenged efforts from domestic and transnational political and economic actors to impose a model of agrarian development based on biotech crops when there were organized peasant movements as well as urban-based agroecological, environmental and consumer rights movements with a critical view towards biotechnology that were ready to combine disruptive repertoires of contention and legal mobilization. This was the case in Brazil. Their mobilization was able to influence the course of technology adoption due to the availability of allies among scientists and political parties, as well as to the opening of media space and access to institutional arenas of policy-making. Different social movement sectors in Brazil saw in the topic of GMOs a possibility for brokerage between different demands and grievances: environmental arguments on ecological risks converged with the discourse of consumer rights; scientific concerns on risks found linkages to the fight for land and peasant rights; and demands for the rule of the law and political participation were aligned with such substantive concerns. Above all, GMOs bridged struggles taking place in the urban centres and the countryside and brought the demands from the agrarian lobby under scrutiny in urban centres such as São Paulo, Porto Alegre, Brasília and Curitiba.

Social mobilization was successful in challenging the institutional and discursive powers of the pro-GMO network by exploring openings in the public debate in the political landscape and in the policy culture with both a risk discourse and a rights-based discourse. However, social mobilization could not counter the material power of biotech firms and agrarian elites in Brazil, who relied on extra-institutional and, often, illegal and violent means to demobilize peasant movements, minimize controversy over GMOs and ensure enough institutional power to allow for their material domination. This was how Brazil became a top producer of GM crops after ten years of controversy. Nevertheless, the discursive and institutional domination of the pro-GMO coalition was never fully accomplished as can be attested to the remaining social mobilization and institutional openings for reigniting controversy at specific moments. Because of the absence of their capacity to persuade others of their world view and for not refraining from using coercion, the pro-biotech actors could not build a situation of hegemony.

Conversely, efforts from the pro-GMO coalition to build a corporate-based, neoliberal, biotech food regime flourished when they were not met with significant social resistance. This was the case of Argentina. Although there were some non-governmental organizations and social movements with a critical view on the effects of spreading GM soy during the 1990s, they were often reactive to the changes by employing defensive strategies to a process that had already gone as far as turning the majority of soy fields into biotech crops. They found no allies to represent their views in the political system or among the scientific community. All this hampered their access to the media and political arenas. Above all, they were not able to establish horizontal linkages between their nodes of resistance. Unconstrained by demands to engage with counter-hegemonic strategies and arguments, the network of actors in favour of biotech crops were able to build a firm hold on state institutions and to have a prominent role in the public discourse, influencing a culturally dominant understanding of the *modelo agrobiotecnológico* to be in the general interest of Argentinean society. It was only with the accumulation of grievances brought by the expansion of GM crops, such as forced evictions and environmental and health damage caused by pesticide spraying, that points of resistance proliferated in the soy-producing regions of Argentina. However, the nodes of resistance have not built enough collective power to challenge the discursive, institutional and material powers that sustain the agrarian model in Argentina.

While relying on a theoretical argument based on five analytical perspectives, the study also contributed to the theoretical debate in each of these areas, thereby advancing it, specifying conditions or pointing to new possibilities. Starting with the macro-sociological theories on the agrarian bases of capitalism, as food regime scholars have noted, the macro-historical agrarian formations are not a direct translation of specific interests but rather the result of social struggles between contending actors, among which social movements are crucial (McMichael 2009). The study confirms the argument from Pechlaner and Otero (2008) that there is no homogenous pattern of expansion of the corporate food regime, as country policies have some leeway in implementing global neoliberal policies and as civil society contributes to shaping the food regime at national and local levels. Again, the lack of appropriate conditions in terms of intellectual property rights or royalties over seeds in Argentina as well as the challenges in the courts against such a system in Brazil showed that countries will not necessarily just follow the conditions that are dictated by the developed countries holding patents on GMOs. Therefore, transgenic seeds did not necessarily overcome barriers to commodification (Kloppenburg 2004) because the application of intellectual property rights is always context-dependent and subject to power disputes.

The research showed how biotechnology contributes to the penetration of capital in agriculture beyond the commodification of seeds. Soy fields in Argentina and in Brazil are an example of how transgenic seeds resistant to pesticides paved the way and catalysed the conversion of agrarian activities of preparing the soil, sowing, managing pests and harvesting into industrial farming, a capital-intensive and chemical-intensive activity. This is why Rauchecker (2015) suggestively speaks of a 'pesticide revolution' in lieu of a transgenic revolution.

This points to another issue of scholarly debate, that is to say, the centrality of biotechnology as a component of the current capitalist agriculture, as claimed by Pechlaner and Otero (2008). Agreeing with the authors by claiming that biotechnology is emblematic of the current food regime, the specific nuance that I would like to add is that the appropriate conditions for the implementation of such a regime lie not necessarily in a strong regime of intellectual property rights over seeds, but much more on a lax regulation of its socio-environmental impacts. The case in Argentina shows that the biotech food regime was installed without the enforcement of intellectual property rights or royalties for GMOs, although this might become now a condition of its maintenance. A lax regulation of health and environmental risks, from the global to the local level, is nevertheless a condition for both the installation and maintenance of the biotech food regime.

These issues also relate to the next body of theoretical literature, namely science and technology studies. All governments are equally responsible for establishing international health and environmental standards for biotechnology and pesticides, and they have proven weak vis-à-vis the global interests of the chemical, seed and food industries. Transnational corporations continue to sway international institutions in promoting a policy culture that supports technology incorporation into food and agriculture while neglecting all evidence of damage. They do so by sticking to a risk assessment that does not hold true outside laboratory conditions. The only challengers to the international safety standards come from the margins of suburban areas in producer countries. This brings to the fore the asymmetries and interconnections between the political economy and the political ecology of GM crops. Trade and retailing nodes of a global commodity chain are able to appropriate rents and 'externalize' socio-environmental costs to the production nodes. Struggles over land, evictions due to pesticide drift, genetic contamination and health impacts of GM fields reveal the challenges for governments in protecting rights in the face of a radicalization of global capitalist dynamics in agrarian activities.

The pesticide issue defies global regulatory institutions in relying on epidemiological data from the rural margins located at the periphery of the chain when establishing safety standards for chemicals and seeds developed in the urban laboratories of the capitalist core economies. In this sense, the present study departs from the argument, defended by Delvenne *et al.* (2013), according to which the regulatory answers to the local problems at the periphery of the global regime might not apply to the core. First, because contaminated communities also exist in developed countries, as the environmental justice activism in the USA has shown (Tesh 2000). Second, and more importantly, evidence of damage to the most vulnerable (often children, elderly and pregnant women) has been the criteria for establishing a safety standard in the core countries where chemical and biotech industry have their headquarters. Proof of health and environmental harm are very difficult to obtain according to the rules of a policy culture based on scientific risk assessment. Therefore, evidence of harm in the most vulnerable groups is often used as a basis from which to extrapolate to all human beings and draw a conclusion. The difference here is that instead of laboratory rats, the cases studied concern entire populations of humans, and their environments.

Therefore, there is enough evidence to argue for the extrapolation of safety standards based on the margins of production in the establishment of truly international safety standards.

The research further adds to cumulative findings on policy culture in technological innovation by showing how regulatory science is used to legitimize GMOs and how the policy culture leaves no room for criticism or recognizes the limits of science. Dissidence was stigmatized, censured and ignored, thereby stalling an argumentative exchange based on scientific methods. The cases present some variations to the argument of robust political cultures in dealing with risks (Jasanoff 2005). By showing how specific networks of actors promote their policy frames, the cases question the idea of a general political culture that results from collective processes of sense-making of a technology. In Argentina, the public was presented with an issue already framed by a powerful network that included the main media groups. In Brazil, there was no robust political culture on technological risks but there was a maintained dispute between two policy frames due to pre-existing differences in how to make sense of technology and agrarian models. This is related to the next issue.

The study has provided further evidence of the importance of structural conditions for public debate and participation to counter influences of economic interests and political power, thereby promoting the observation of human and constitutional rights. This became clear in the divergence between Argentina and Brazil. First, the institutional mechanisms of participatory democracy established in Brazil following the adoption of the federal constitution in 1988 envision that representatives from civil society are invited to participate as human rights defenders owing to their reputation and traditions, in addition to their expertise in the issue area. Second, by allowing civil society participation, there is an opening for rights claims, which can dispute influence with knowledge claims in the struggles over the public meaning of a given policy. Related to that is a third factor: the existence of plural media coverage. There are more chances that a public policy that is not publicly defined as technocratic but instead relates to wider societal interest and rights will receive media coverage, particularly in newspaper sections concerning politics and society and not only innovation and economy. Another factor explaining media coverage was the commercial media alignments (Hallin and Mancini 2004). In Brazil, the main media groups are more urban and industry based, while in Argentina they have high representation in agribusiness.

The research highlights the importance of timing among the appropriate conditions for social mobilization concerning GMOs. While systemic imperatives in the case of innovation policy include a fast launching of products (Motta 2008), this poses an obstacle for processes of meaning-making and opinion formation. Democratic theory provides a normative basis as well as analytical distinctions in order to identify if a public policy undergoes legitimizing processes in which public opinion formation takes place in the public sphere (Habermas 1992, 2008). In this case, a policy must be supported by arguments that give reasons for its adoption and respond to raised objections. It allows for the identification of illegitimate uses of socioeconomic and political power, having as a consequence a public policy with gaps in legitimacy. The study has shown that the latter prevailed in both cases. In order to describe how this occurred, the research moved from the macro-level of theorizing to the micro- and meso-level of actors as well as their networks and interactions. The theoretical argument of the book contributes to the theory and research on the structural conditions for deliberative democracy by bringing into the analysis the actor constellation that employed strategies to avoid that GM crops become objects of wide processes of opinion and will formation while hindering the role of social movements in making GMOs a political problem.

In order to analyse how the coalition of biotech corporations, experts, policy-makers and agrarian elites acted to promote a political consensus in favour of GMOs, the research used the concept of bio-hegemony, coined by Newell (2009). In both countries, the fact that networks of actors related to large-scale capitalist farming and to chemical industries were the main promoters of GM crops casts doubt on the pro-poor and sustainability rhetoric with which they promoted GMOs. The case of Brazil highlights how material power depends on the other dimensions to expand since the legal moratorium and uncertain legal status of GM soy were obstacles to its adoption. But, at the same time, it also shows how the pro-GMO coalition was ready to use illegality to increase institutional power. Indeed, the strategy of *fait accompli* was key in the conversion of Brazil into a top producer of biotech crops. In Argentina, the widespread adoption of GM seeds without royalty payments ensured the material domination of biotech crops (although it is not possible to know whether this was a strategy from the technology holder or was unintended) and then provided the basis for demanding an institutional arrangement for royalty collection. Both cases show the existence of rifts inside this coalition regarding the distribution of rents. These rifts took place between Monsanto and farmers considering royalty payments in both countries, and in Argentina the main rift was between the state and farmers regarding taxes.

Shortcomings of Newell's usage of bio-hegemony also become clear in the cases studied. Coming back to Gramsci's (1971) theorization of hegemony and symbolic domination in capitalism, the starting point is that coercion alone was not enough to maintain the dominance of the material relations of production as the proletariat could easily outnumber the dominant class and, in turn, revolt. Therefore, for Gramsci there was also domination in the superstructure of society, that is to say, the ideology in which most proletariat shared the values of the dominant class and aspired to ascend socially and become part of it. However, some problems arise when applying the concept to characterize the adoption of GM crops in Argentina and Brazil.

The reconstruction of almost two decades of the accounts of social movements have shown that, first, there has been mounting mobilization against the agrarian model from the beginning of its expansion and, second, such mobilization has been silenced by coercion and violence. Even if one can speak of bio-hegemony in the sense that the general society is neither aware of the conflicts nor mobilized by the issue, it cannot be said that it is undisputed due to the existence of various social movements carrying contentious meanings about the 'transgenic revolution' and, above all, because the network of pro-GMO actors cannot deflect critiques without refraining from coercion. In addition, the absence of mobilization does not mean that there is symbolic domination, as James Scott (1987) showed in an exemplary study among peasants in Malaysia. Indeed, in the cases studied, although many had to live with GM crops, it was not with their consent and rather contrary to their wishes (Lapegna 2015). Finally, in Argentina and Brazil, the legitimacy of the agrarian model became increasingly dependent on that of a concept of development based on economic growth combined with social policies of poverty reduction, if not patronage. Social movements whose grass-roots bases are recipients of social programmes claim, however, to prefer not being dependent on such programmes and instead rely on state policies that support their agrarian activities and protect them from the expansion of agribusiness.

This brings us to the value added by peasant studies in understanding social disputes over GM crops, in shedding light on the new political subjects who strengthened their mobilization during the last decades (Borras *et al.* 2008). Contemporary peasant movements contradict the expectations of most social theories that rely on the assumptions of modernization theory; this equates (capitalist) development with economic growth, industrialization, rural outmigration and urbanization. Consequently, social theory, liberal and Marxist alike, has contributed to the view of peasants as pre-modern and deemed to disappear (McMichael 2008). Yet, the renewal of agrarian movements at the turn of the century has shown that not only are they disputing national processes of agrarian development but they are also constituting an important transnational node of resistance against the current formation of global agrarian capitalism. They represent a critique of the narrative of agrarian modernization, which promises social mobility to the rural poor through market inclusion. In contrast to such residual theories of agrarian change (Borras 2009), which functioned as powerful capitalist ideologies fuelling the adoption of GMOs among small farmers in both countries, peasant movements construct a political identity based on class struggles (Petras and Veltmeyer 2001).

In contrast to scholars who defend the inclusion of the rural poor in the markets through the adoption of GMOs and the agribusiness model (Herring 2007; Paarlberg 2000), this study provided evidence that the rural poor want to be included in the welfare state – in addition to receive a minimum income – and to have proper agrarian policies to support their way of producing and their livelihoods. Many small farmers that have adopted the discourse of technological modernization in Argentina later became critical of it owing to the fact that they lost their lands. In Brazil, most peasants who adopted GM crops have regretted it because they could not afford the increasing costs of pest resistance and were directly affected by the high levels of pesticide required for crop management.

The book revealed the importance of alliances between urban NGOs, often with professional backgrounds in law, agronomy and communication, and agrarian movements in order to challenge the biotech food regime. While the former often had more resources to engage in lobbying and in legal mobilization, as well as to mobilize expertise, the latter brought a high mobilization power to conduct mass disruptive actions. Their alliances bridged the gap between two debates: biotechnology and agrarian policy. In this sense, this research reaches a conclusion that coincides with Schurman and Munro (2010), that is to say, that activists in the USA, despite being very organized and professional, failed in reaching a grass-roots base, which was an obstacle for that overall fight. Indeed, the presence of organized mass movements of the agrarian poor in Brazil was fundamental to widen the debate on rights claims. On the other hand, the Argentinean process of grass-roots mobilization has not consolidated a strong alliance with professionalized NGOs and urban social movements that could monitor and follow national policy-making on GMOs in different arenas.

The cases showed that reliance on scientific counter-expertise and disputing official risk assessments (Arancibia 2013; Kinchy 2010) is a path of argumentation that movements do not avoid when contesting GM crops. However, their knowledge claims are often stigmatized and not taken seriously. Associating risk discourse with rights claims was a more influential framing strategy. In Brazil, partial victories are attributable to appeals to democratic rights and the respect for legal provisions but not on knowledge claims. Local victories in Argentina stem from justice claims that seek to stop violations of rights rather than a new consensual scientific knowledge.

Having said this, it cannot be stated, based on these cases, that the demands for rights and justice were reflected in policy changes in the direction of health and environmental protection or of a stronger support for an alternative agrarian model oriented towards achieving food sovereignty, the master frame of peasant movements. In this sense, while the study verifies the conclusions from past research (Newell 2008), it also adds some specific assessments. In the debate on biotechnology in Brazil, social participation in the legislative and expert arenas was fundamental in establishing legal and institutional room for public scrutiny and the precautionary principle. This result can serve future opportunities for activism and, not less important, for attributing responsibility for damage. Social mobilization contributed to drawing a line between GMOs and non-GMOs, whereby making the former a controversial and politicized topic. This has paved the way for alliances with peasant movements. The debate on agrarian policy, however, cannot be considered as having a very positive balance for peasant movements, which strive to change the agrarian model. However, their sheer existence forms a counter-hegemonic agency that manifests alternatives to the dominant model of agrarian development.

Here, a short note on social movement outcomes is necessary. The last two paragraphs focused on policy change in the regulation of GMOs and agrarian policies as a way to measure the success of a movement. But the research showed how social movements' outcomes are invisible when the indicator is the position of these countries as top GM producers. By reconstructing the downside of this process of agrarian change, the book narrated the formation of networks and solidarity among activists, the building of new political identities and the development of a critical interpretation to the narratives of technological modernization.

In addition to understanding social mobilization, it is necessary to be able to explain the processes of demobilization (Tarrow 2011). This study makes a contribution in this direction, taking into account in particular the reactions from elites and authorities to social mobilization. Movements fighting GM crops in Brazil and Argentina demanded a fundamental change in the agrarian model of these countries, which in turn jeopardized the interests of powerful actors. Social movement scholars have argued (Gamson 1992) that in such cases suppression is much more likely than in the case of movements demanding smaller reforms. In fact, in both countries, activists were victims of violence and murder conducted or mandated by actors from the bio-hegemonic network as well as of persecution and criminalization. Their civil and political rights were violated; the act of violence against an activist, who is a human rights defender, is an exemplary act to communicate that political activism and mobilization will not be tolerated. The state selectively applied the law to threaten activists and the rural poor, and to withhold its enforcement when these were the victims. By such a selection of who has what types of rights, the state has promoted the use of violence in the countryside as a mechanism to advance the biotech food regime. Other mechanisms of demobilization at play among the rural poor were poverty alleviation schemes and the adoption of GMOs among the bases of agrarian movements, as already mentioned.

Last but not least, the present research showed how gender, despite not being taken into account by the theories explaining social disputes over GMOs, is a key category. The research in Argentina showed how women living in poor neighbour-hoods in the transition between urban and rural zones experienced the effects of the expansion of GM fields in a specific manner. They rejected fatalist explanations for the health problems suffered in their families and community and collectively searched for causes. They created powerful collective action frames. It was a specific gendered con-text in which mobilization took place, where women are ascribed roles in the realm of reproduction and care. These conditions and relations shaped their political participa-tion. From their traditional (and socially constructed) roles as mothers, women living close to GM soy fields created their own political culture of citizenship (Schild 1994).

However, they went much beyond the traditionally ascribed roles of caregivers and were able to politicize an issue that was otherwise taboo in Argentina, due to the hegemonic view of agrarian activities as representing the general interests of soci-ety. They were able to create frictions in this hegemonic understanding from their marginal position in the political realm. In Argentinean history, the dictatorship had policitized women as the state violated the rights of all; now, agrarian activities, when threatening life and violating rights, had also politicized women. Fighting the *modelo agrobiotecnológico*, women acted not as 'women *without* rights' that fight to obtain women rights, but rather as 'women *for* and *in relation* to human rights', namely fight-ing for the universal right to have rights and to participate in the definition of what are human rights in a specific context and time (Jelin 1996, 194). In short, this study contributed to the theoretical debate on mobilization in this issue area by identify-ing processes in which the identity as women has been crucial for social mobilization against GM crops.

Beyond GMOs: social movements, democracy and capitalism

The book strived to make a contribution that goes beyond the issue area of GMOs and relates to wider debates on the role of social mobilization in struggles for dem-ocracy and rights protection vis-à-vis market logics and political interests. In more general terms, the study addressed the question on the conditions in which social movements and civil society organizations challenge the script of 'business as usual' in public policies, that is to say, when their claims for rights and their action repertoire have the power to counteract the interests of other actors, such as networks driven by market interests that often are able to influence state policies in their favour. Just as important is to understand the obstacles to mobilization such as the conditions that suppress and silence dissent. Such endeavour requires considering national political contexts in their relations to global capitalism.

The debate on the relationships between democracy and capitalism has become prominent again, in particular after the global financial crisis (Streeck 2011; Wagner 2013). Citizens dispute the political influence over public policies with global corpora-tions that have often greater revenues than countries, be they biotechnological or seed firms, as in the case at hand, or financial institutions and banks, as the current wave of protests from the *indignados* (the outraged) and the occupy movement attests. The comparison of cases in which state responses varied, which is related to civil society mobilization, shows that there is room for politics and for different state responses vis-à-vis the pressures of global markets. This was the point of departure of this study, and it is relevant to understand the room for politics in a moment in which the zeitgeist seems to be giving in to gloomy diagnoses that the role of the state in contemporary capitalism is to administer financial markets.

Unfortunately, the stories narrated in this book give little room for an enthusiastic assessment that social mobilization was able to deter the socio-environmental negative effects of the expansion of global agrarian capitalism with the introduction of GM seeds, even less to say that they were able to promote an alternative agrarian future to the one characterized by the constitution of a corporate and neoliberal food regime.

However, the initial assumption that contestation is necessary for emancipatory social change has indeed been reinforced, at least in its negative form: without social mobilization, global agrarian capitalism brings with it more inequalities and the diminishing of rights. Rather, the exercise is to think what would have happened in the absence of social mobilization. The absence of an initial organized mobilization in Argentina was crucial for a rapid and strong consolidation of agribusiness, with a much more radical and abrupt rate of deforestation, chemical contamination, forced evictions and expansion of the agrarian border. The existence of social movement organizations and networks in Brazil was fundamental in reacting and critically disputing the public legitimacy and policy on GMOs. As argued by Schurman and Munro (2010), the most important impact of activism was to construct the categorical difference between GMOs and non-GMOs, and, although it could not halt biotech crops, the movement affected the industry's plans by making GMOs a politicized and contested technology.

Extrapolating to other cases, social mobilization can be the main agency behind transforming an issue into a political problem, as already stated by social movement scholars (Rucht 1994). This proposition relativizes the role ascribed to media in democracies (Habermas 1992, 2008). The case studies showed that commercial alignments in the media were key to portraying the arrival of biotechnology as agrarian innovation and sustaining the preferred framing from the pro-GMO coalition. It was social mobilization that brought the issue to the media agenda in Brazil, thereby influencing media discourse and bringing into play different actors and policy frames. Therefore, social movements and affected people can be transformed into the main agency to bring or counter new issues in the political agenda. As the study has revealed, social mobilization depends on the existence of organizational bases and networks, on the construction of contentious meanings about a new issue, on a combination of disruptive and institutionalized action repertoire, on horizontal linkages among social movements, and on the existence of allies in the political and scientific systems.

A major factor influencing the opportunities of movements was timing. In Brazil, it affected the trajectory of adoption of GM crops from the beginning to the end, as activists created opportunities for their continuous engagement, such as participatory and transparency provisions in the law. Having led the path, activists, even when faced with a closed policy, could nevertheless revive the issue at different times, when there was a catalytic event, and activate networks for a joint action. The relevance of this finding is that it is much harder for a social movement to challenge the state of affairs in comparison to the capacity of a coalition of economic interests to do so and achieve their favoured outcome. Democracy under capitalism is more prone to the closure of societal influence rather than to deflecting pressures from the economic lobby. Therefore, timing in leading the path is key to influencing the political treatment of an issue.

In order for social movements to be able to lead, some previous conditions are necessary. These relate to the structure of political opportunities, such as access to institutional arenas of decision-making and access to media. The list here is not exhaustive; to the contrary, it points to possibilities for further research. A first condition is the existence of participatory mechanisms in all policy areas. This is how civil society knows about the political agenda, is able to learn about new topics, constructs its critical interpretations and defines its positions. In the absence of public participation, the pattern is that national and international private corporations, experts and policy-makers dominate policy.[1] In short, participatory democracy is an important agenda for continuous research.[2]

A second condition is independence and plurality of media (Habermas 2008) as discursive opportunities are crucial to social movements. In both countries, media reforms are high on the political agenda (Mauersberger 2015). In Argentina, although a new media bill passed, the government is having a hard time implementing it and media groups continue to be very concentrated. In Brazil, after a right-wing weekly magazine tried to construct a media scandal two days before the 2014 presidential elections, relying on false testimonies, the re-elected president mentioned in her inaugural address that she would make a media reform a part of her new mandate. Whether these reforms will advance and whose interests they will serve are also issues that will demand continuous research.

A third condition favourable to societal influence is the division of power among the executive, legislative and judicial branches. This increases the opportunities of civil society to influence political outcomes, demand rights as well as insist on the monitoring and control of the other powers. Research agendas related to this factor are the debates on law and social mobilization (Cardoso and Fanti 2013; Hilson 2002; McCann 2006; Rodriguez 2013) as well as law and politics (Nobre and Rodriguez 2011) and also on the relations between the parliament and the executive and how they affect social movements.

However, social contestation is not enough, which brings us to the conditions for social movements to remain mobilized. It is necessary that social mobilization is free from constraints. In both countries, activists were faced with a threatening context for mobilization efforts, in particular due to the politics of 'care and punishment' (Schild 2013). This took the form of providing cash transfers in order for the poor to take care of themselves and criminalize poverty and dissent. Therefore, a fourth factor is protection from violence and processes of criminalization from which activists and the rural (and urban) poor are victims. Research on the relationships between inequalities and violence, both in the city and countryside (Kay 2001), are also relevant to the inquiry into the emancipatory possibilities of social mobilization.

Maintained mobilization is conditioned on a fifth factor, that is to say, protection against patronage and political co-optation. Repeatedly, the social movements analysed were co-opted by new governments with a progressive agenda, or had their bases demobilized due to the effects of poverty reduction schemes. For movement leaders, however, the possibility of weakening the government and giving room to the opposition to win political space was perceived as something to be avoided at the cost of abandoning some of their claims or at least forego disruptive strategies to achieve them. Thus, they have oriented their bases to vote to re-elect the incumbent government due to fear of becoming again invisible to the state. As pointed out by Gamson (1992), social movements might not seek more advantages but increased access, expecting that access to institutional spaces in the policy process becomes a durable opportunity structure for the movement (Tarrow 2005).

The book highlighted the difficulties faced by social movements in contexts of high social inequality, in which they see themselves forced to reduce their demands on the state to a minimum shared consensus of reducing poverty. All agree with the state provision of social policy but seem to share the interpretation that this might be detrimental for mobilization and, more importantly, they would prefer having rights guaranteed than be compensated in form of cash transfers. For them, political opposition could mean even a regression. However, 'the compensatory state' (Gudynas 2012) appears to be hegemonic in Latin America: just as much as the extraction of natural resources, cash transfer programmes have been institutionalized as a state policy that no government, from the right or the left, will refrain from (Lavinas 2013). The relation between social movements and the new left remains an open and important research agenda (Bringel and Falero 2014; Prevost *et al.* 2012) together with the relation between social mobilization, poverty and patronage politics (Lapegna 2013b, 2015).

These research agendas hint at the crucial role of capitalism in social mobilization and demobilization. The book highlighted the need to embed political opportunity structures in global and national markets, which could be translated for social movement studies as a 'political-economic opportunity structure' (Motta 2015). The study showed how different interpretations on whether the primacy lies in politics or markets have influenced social movement struggles. Political economy matters for social movements both at the national and at the global levels.

By considering the national political economy as part of the context conditions for movement struggles, this study looked not only at changes in political alliances but also at how the correlation of forces is affected by economic conditions. In the cases studied, the power of agrarian exporters in the polity increased as commodity prices peaked. Following their election with promises of a progressive agenda, leftist governments promoted commodities for export as an important source of revenues and justified with a stronger role of the state in ensuring a combination of economic growth and poverty reduction. The model had good results with the inclusion of many in the labour and consumption market; it created mass consumption societies. This national context showed how capitalist ideologies of the citizen as a consumer are key for legitimizing the commodity-based model. It constructs a gap between grievances in the places where extractive industries take place and the relative affluence in the urban centres. While for governments this was an opportunity to construct complementarities between commodity-based growth and mass political support, for social movements it represented a challenge in deploying brokerage mechanisms between rural and urban struggles.

The model, however, is reaching its limits, as evidenced by the mounting socio-environmental mobilizations in both countries. The political ecology of the relations between commodity based-growth and socio-environmental conflicts is a vivid research area worldwide and in Latin America, in particular, with different conceptualizations such as 'place-based struggles', 'environmental justice', 'neo-extractivism' and 'eco-territorial turn' (Burchardt *et al.* 2013; Castro and Motta 2015; Escobar 2001, 2008; Gudynas 2012; Johnson and Bebbington 2011; Seoane *et al.* 2006; Svampa 2012; Zhouri and Oliveira 2012). Furthermore, the wide range of demands from civil society during the June 2013 protests in Brazil demonstrated that consumption is no substitute for rights. Beyond inclusion in the market, citizens are demanding, among others, inclusion in the welfare state with adequate health and education services, the right to mobility and to sustainable cities, and the right to participate in the decisions affecting their lives. Until now, the government has failed to respond to the increasing demands for rights. It remains to be investigated how the decrease in economic growth and commodity prices will affect the national political economy of the region: will the protected class struggle come to the fore as governments will have to make political choices between a progressive tax reform or cutting social expenditure? Recent developments in Brazil show how in times of economic and political crisis taxes are placed on the agenda but not necessarily in progressive ways; the past example of Argentina revealed how state responses to crisis can increase inequality (Valdés forthcoming). For social mobilization, the main task is to make a bridge between territorially based resistances – such as those from peasants, indigenous communities and sprayed neighbourhoods – and urban struggles.

Last but not least, the global political economy in each issue area is a relevant context condition to understand the possibilities of social mobilization. In the case at hand, the concept of the global commodity chain allowed the national playing fields to be situated in the world agrifood system, taking into account its asymmetries. This can be applied to other global commodity chains in extractive industries such agrarian and forest products, minerals and energy as well as industrial goods such as clothing – an issue of great contention nowadays. This paves the way for asking such questions as what are the impacts of consumer activism on the level of production and whether and how consumer activism contributes to or curtails social struggles locally. The concept highlights the importance of alliances across the globe among nodes of the chain.

For instance, the struggle of global anti-GMO movements and of each national movement against biotechnology – in particular those taking place in import markets – will only be effective if linked to struggles taking place among peasants and the rural poor. The conclusion from the cases studied is that it not enough for social mobilization to win a national or a regional policy, like in the EU, where no GM crops are cultivated and Europeans import GM crops produced elsewhere for processed food or animal farming. As long as the EU relies on a global commodity chain for grains and vegetable protein, it contributes to the asymmetrical distribution of risks and human rights violations of this chain to the soy-producing regions. In addition to an alliance between urban and rural movements in each country, only a transnational solidarity network, fighting simultaneously in many countries, might successfully influence the shaping of the global food regime. Global ethnography (Burawoy 2000) offers a good promise to address such issues.

The global food, energy, environmental and financial crises have placed the agrifood system on the political agenda, where agrarian futures are under contestation. Social contestation might be the main agency for bringing rights-based arguments to the debate to dispute influence with economic and political interests in the shaping of global patterns of food and agrarian production and distribution. Social movements emphasize the difference between commodities and food as well as the distance between agrarian production and food consumption, which is mediated by sociopolitical contexts. This means that the main factor influencing the capacity of the world in feeding its population is not volume of production and technological productivity but unequal conditions of access to food, which are a sociopolitical problem.

The case studies have also provided a more nuanced view of the efficiency from the so-called breadbasket countries. Their higher competitiveness is attributed to the absence of state subsidies as compared to developing countries. The study showed that the increased volume of grain production depended on other forms of subsidies, such as amnesty for environmental crimes and human rights violations that have accompanied the expansion of GM crops in Argentina and Brazil. Eschewing democratic processes was also crucial to the advancement of the model. This sheds doubt on the proclaimed technological efficiency of biotechnology in a global dispute over agrarian futures. Therefore, the argument of the book aspires to provide some food for thought for those engaged in such global debates: the expansion of GM crops in Argentina and Brazil was not an automatic consequence of its technological superiority. As the stories revealed, it resulted from asymmetrical social disputes over state policies, and, ultimately, from political decisions. It was only possible at high costs for the environment, human rights and democracy. In short, the advancement of a world agrifood system – based on the commodification of biotech seeds, large-scale chemical agriculture, industrial farming and processed foods – is therefore a politically mediated development as it cannot be furthered without the active engagement of states and a network of actors that combine strategies of persuasion and coercion. Nevertheless, with the absence of social mobilization, such developments have worse consequences for the exercise of democratic rights. The latter are not an automatic result of legal frameworks as the social movements protagonists of these stories warned.

Notes

1 Recent events in Brazil reinforce the power of public participation, as the legislative branch rejected the president's proposal of holding a referendum for a political reform in response to the June 2013 protests. In October 2014, the same legislative body rejected a decree that established a national system of public participation. The decree only organized previously existing bodies but nevertheless frightened the members of the National Congress. These moves are indicative of the fears that dominant elites have of the public and of losing privileges, but they also sign that such privileges are under critique.

2 Note that public consultations for achieving prior informed consent from populations locally affected by development projects (as foreseen by the International Labour Organization (ILO) Convention 169) have been propagated worldwide by international organizations as a solution to socio-environmental conflicts. However, they fall short of many requirements of being seen as legitimate ways of public participation, not least because usually they are put in place when decisions have already been taken. A show of distrust in such mechanisms was given by indigenous movements in Brazil, which have abandoned the process of regulation of the ILO 169.

Appendix: list of interviews

All interviews conducted by the author personally, except when stated otherwise

Advisory and Services for Projects in Alternative Agriculture, AS–PTA (Rio de Janeiro, 13 February 2012).

Aranda, Darío. Journalist, *Página/12* (Buenos Aires, 6 August 2012).

Ávila, Medardo. Former Secretary of Health Córdoba and member from Doctors from Sprayed People and Collective Stop Spraying Córdoba (Córdoba, 8 August 2012).

Brazilian Institute for Consumer Protection, IDEC (São Paulo, via Skype, 7 March 2013).

Carrasco, Andrés. Scientist, University of Buenos Aires and National Council for Scientific and Technical Research, CONICET (Buenos Aires, 10 August 2012).

Córdoba Peasant Movement, MCC (Universidad Campesina, Ojos de Água, Provincia de Santiago de Estero, 21 and 22 February 2013).

Genetic Resources Action International (GRAIN) Argentina (Buenos Aires, via Skype, 5 September 2013).

Greenpeace Argentina (Buenos Aires, 30 July 2012).

Greenpeace Brazil (São Paulo, 3 February 2012).

Group of Rural Reflexion (GRR) (Merlo, Buenos Aires, 6 August 2012).

Former official from the Brazilian Institute for Consumer Protection, IDEC (São Paulo, via Skype, 5 March 2013).

Former campaigner Greenpeace Argentina (Buenos Aires, 9 August 2012).

Kageyama, Paulo. Scientist, Universidade de São Paulo; former member of CTNBio, (São Paulo, via Skype, 10 July 2013).

Luque, Stella. Scientist, Faculty of Agrarian Sciences and activist from Collective Stop Spraying Córdoba (Córdoba, 8 August 2012).

Mothers of Ituzaingó Anexo (Córdoba, 7 and 8 August 2012).

Movement of Landless Workers, MST (Brasília, via Skype, 12 July 2013).

Movement of Small Farmers, MPA (São Paulo, 24 January 2012).

National Network of Ecological Action, RENACE (Buenos Aires, 10 August 2012).

Nodari, Rubens. Scientist, Federal University of Santa Catarina; former Director of Ministry of Environment and former member of CTNBio (Santa Catarina, via Skype, 2 July 2013).

Santiago de Estero Peasant Movement-Vía Campesina, MOCASE-VC (Buenos Aires, 16 February 2013).

Santiago de Estero Peasant Movement-Vía Campesina, MOCASE-VC (Universidad Campesina, Ojos de Água, Provincia de Santiago de Estero, 21 February 2013).

Seeds from the South, NGO part of Collective Stop Spraying Córdoba (Córdoba, 8 August 2012).

Service to the Popular Culture, SERCUPO (Monte Grande, Provincia de Buenos Aires, 18 February 2013).

Terra de Direitos (Curitiba, via Skype, 7 March 2013).

References

Aiuto, Maria Ines. 2006. 'Pueblos Fumigados. Informe Sobre la Problemática del Uso de Plaguicidas en las Principales Provincias Sojeras de la Argentina'. Grupo de Reflexión Rural. www.grr.org.ar/campanapdf/.

——. 2009. 'Pueblos Fumigados. Informe Sobre la Problemática del Uso de Plaguicidas en las Principales Provincias Sojeras de la Argentina'. Grupo de Reflexión Rural. www.grr.org.ar/campanapdf/.

Anderson, Paul Nicholas. 2004. 'What Rights Are Eclipsed When Risk Is Defined by Corporatism?: Governance and GM Food'. *Theory, Culture & Society* 21 (6): 155–70.

ANEC. 2014. 'Farelo e Óleo de Soja'. Accessed 3 March. www.anec.com.br/pdf/ExportacaUltimos23anosFareloDeSoja.pdf.

Arancibia, Florencia. 2013. 'Challenging the Bioeconomy: The Dynamics of Collective Action in Argentina'. *Technology in Society* 35 (2): 79–92.

Aranda, Darío. 2009. 'El Tóxico de los Campos'. *Página12*, 13 April. www.pagina12.com.ar/diario/elpais/1-123111-2009-04-13.html.

——. 2010. 'Otra Campaña del Desierto, Ahora por la Soja'. *Página12*, 12 October. www.pagina12.com.ar/diario/elpais/1-154770-2010-10-12.html.

——. 2013. 'El Árbol y el Bosque'. *Lavaca*, March.

Aronskind, Ricardo, and Gabriel Vommaro, eds. 2010. *Campos de Batalla: Las Rutas, los Medios y las Plazas en el Nuevo Conflicto Agrario*. Buenos Aires: Prometeo Libros.

Arza, Valeria, María Eugenia Fazio, Laura Goldberg, and Patrick van Zwanenberg. 2010. 'Problemas de la Regulación en Semillas: El Caso del Algodón Transgénico en el Chaco'. *Desarrollo Económico* 49 (196): 605–28.

Auyero, Javier. 2004. 'The Moral Politics of Argentine Crowds'. *Mobilization* 9 (3): 311–26.

Auyero, Javier, and Timothy Patrick Moran. 2007. 'The Dynamics of Collective Violence: Dissecting Food Riots in Contemporary Argentina'. *Social Forces* 85 (3): 1341–67.

Baptista, Renata, and Mari Tortato. 2006. 'Via Campesina Invade Multinacional'. *Folha de São Paulo*, 15 March.

Barbetta, Pablo. 2009. 'El Derecho Distorsionado: Una Interpretación de los Desalojos Campesinos Desde el Análisis del Campo Jurídico'. In *La Argentina Rural: De la Agricultura Familiar a los Agronegocios*, edited by Carla Gras, Valéria A. Hernández, and Christophe Albaladejo, 237–56. Buenos Aires: Biblos.

Bardocz, Susan, Ann Clark, Stanley Ewen, Michael Hansen, Jack A. Heinemann, Jonathan Latham, Arpad Pusztai, David Schubert, and Allison Wilson. 2012. 'Seralini and Science: An Open Letter'. *Independent Science News*. Accessed 2 October.

http://independentsciencenews.org/health/seralini-and-science-nk603-rat-study-roundup/.

Barker, Colin, Laurence Cox, John Krinsky, and Alf Gunvald, eds. 2013. *Marxism and Social Movements*. Vol. 46. Leiden: Brill.

Barros e Silva, Fernando. 2001. 'Justica Deixa Bové Ficar; Governo Recorre'. *Folha de São Paulo*, 31 January.

Bauer, Martin W. 2005. 'Distinguishing Red and Green Biotechnology: Cultivation Effects of the Elite Press'. *International Journal of Public Opinion Research* 17 (1): 63–89.

Beck, Ulrich. 1986. *Risikogesellschaft: Auf dem Weg in eine Andere Moderne*. Frankfurt am Main: Suhrkamp Verlag.

———. 2007. *Weltrisikogesellschaft: Auf der Suche Nach der Verlorenen Sicherheit*. 1st ed. Frankfurt am Main: Suhrkamp Verlag.

Benford, Robert D. 1993. 'Frame Disputes Within the Nuclear Disarmament Movement'. *Social Forces* 71 (3): 677–701.

Benford, Robert D., and David A. Snow. 2000. 'Framing Processes and Social Movements: An Overview and Assessment'. *Annual Review of Sociology* 26 (January): 611–39.

Benthien, Patrícia Faraco. 2010. 'Transgenia Agrícola e Modernidade: Um Estudo Sobre o Processo de Inserção Comercial de Sementes Transgênicas Nas Sociedades Brasileira e Argentina a Partir dos Anos 1990'. PhD Thesis, Universidade Estadual de Campinas.

Berger, Mauricio. 2013. *Cuerpo. Experiencia. Narración. Autoorganización Ciudadana en Situaciones de Contaminación Ambiental*. Córdoba, Argentina: Ediciones del Boulevard.

Berg-Schlosser, Dirk, and Gisèle De Meur. 2009. 'Comparative Research Design: Case and Variable Selection'. In *Configurational Comparative Methods: Qualitative Comparative Analysis (QCA) and Related Techniques*, edited by Benoît Rihoux and Charles C. Ragin, 19–32. Thousand Oaks, CA: SAGE.

Bernstein, Henry. 2014. 'Food Sovereignty via the "Peasant Way": A Sceptical View'. *Journal of Peasant Studies*, 41 (6): 1031–1063.

Bidaseca, Karina. 2003. 'El Movimiento de Mujeres Agropecuarias en Luchas: Acciones Colectivas y Alianzas Transnacionales'. In *Más Allá de la Nación: Las Escalas Múltiples de los Movimientos Sociales*, edited by Elizabeth Jelin, 161–202. Buenos Aires: Libros del Zorzal.

Bisang, Roberto, Guillermo Anlló, and Mercedes Campi. 2008. 'Una Revolución (no Tan) Silenciosa. Claves Para Repensar el Agro en Argentina'. *Desarrollo Económico* 48 (190/191): 165–207.

Bissoli, Luiza Duarte. 2012. 'Ação Judicial: Uma Estratégia da Sociedade Civil Contra o Fomento dos Transgênicos'. In *Anais do 36° Encontro Anual da ANPOCS*. Caxambu, Minas Gerais, Brazil.

———. 2013. 'Ativismo Judicial Nas Lutas da Sociedade Civil Contra os Transgênicos'. *Primeiros Estudos* 4 (June): 34–45.

Bonner, Michelle D. 2007. *Sustaining Human Rights: Women and Argentine Human Rights Organizations*. University Park, PA: The Pennsylvania State University Press.

Bonneuil, Christophe, and Les Levidow. 2012. 'How Does the World Trade Organization Know? The Mobilization and Staging of Scientific Expertise in the GMO Trade Dispute'. *Social Studies of Science* 42 (1): 75–100.

Borras, Saturnino M. 2009. 'Agrarian Change and Peasant Studies: Changes, Continuities and Challenges – an Introduction'. *Journal of Peasant Studies* 36 (1): 5–31.

Borras, Saturnino M., Marc Edelman, Cristóbal Kay, Saturnino M. Borras Jr, Canada Research Chair, R. C. Edelmanessor, and Cristóbal Kayessor, eds. 2008. *Transnational Agrarian Movements Confronting Globalization*. Chichester: Wiley-Blackwell.

Bottaro, Lorena, and Marian Sola Álvarez. 2012. 'Acción Colectiva y Ampliación de Demandas Luego de la Crisis de 2001: Las Particularidades de los Movimientos Socioambientales'. In *Problemas Socioeconómicos de la Argentina Contemporánea, 1976–2010*, edited by Mariana Luzzi, 1st ed., 401–20. Los Polvorines: Univ. Nacional de General Sarmiento.

Boy, Daniel, Dominique Donnet Kamel, and Philippe Roqueplo. 2000. 'Un Exemple de Démocratie Participative : La « Conférence de Citoyens » sur les Organismes Génétiquement Modifiés'. *Revue Française de Science Politique* 50 (4): 779–810.

Bresser-Pereira, Luiz Carlos. 2012. 'Structuralist Macroeconomics and the New Developmentalism'. *Revista de Economia Política* 32 (3): 347–66.

Bringel, Breno, and Alfredo Falero. 2014. 'Movimientos Sociales y Gobiernos en América Latina: Nuevos Escenarios, Tipología de Relaciones y Formas Estado/ Movimento'. *Cadernos de Trabalho Netsal* 2 (5): 1–23.

Bröer, Christian, and Jan Willen Duyvendak. 2009. 'Discursive Opportunities, Feeling Rules, and the Rise of Protests Against Aircraft Noise'. *Mobilization* 14 (3): 337–56.

Burachik, Moisés, and Patricia Traynor. 2002. 'Analysis of a National Biosafety System: Regulatory Policies and Procedures in Argentina'. ISNAR Country Report 63. The Hague: International Service for National Agricultural Research.

Burawoy, Michael. 2000. *Global Ethnography: Forces, Connections, and Imaginations in a Postmodern World*. Berkeley, CA: University of California Press.

Burchardt, Hans-Jürgen, Kristina Dietz, and Rainer Öhlschläger, eds. 2013. *Umwelt und Entwicklung im 21. Jahrhundert: Impulse und Analysen aus Lateinamerika*. Baden-Baden: Nomos Verlagsgesellschaft.

Cáceres, Daniel M. 2015. 'Accumulation by Dispossession and Socio-Environmental Conflicts Caused by the Expansion of Agribusiness in Argentina'. *Journal of Agrarian Change* 15 (1): 116–47.

Calhoun, Craig, ed. 1992. *Habermas and the Public Sphere*. Cambridge, MA: The MIT Press.

Camandone, Julietta. 2014. 'El Gobierno Argentino Contra la Soja Ronaldinho'. *El Cronista Comercial*, 4 September. www.cronista.com/negocios/El-gobierno-argentino-contra-la-soja-Ronaldinho-20110902-0152.html.

Camara, Maria Clara Coelho. 2011. 'Regulamentação e Atuação do Governo e do Congresso Nacional Sobre os Alimentos Transgênicos no Brasil: Uma Questão de (In)segurança Alimentar'. PhD Thesis, Escola Nacional de Saúde Pública Sergio Arouca.

Campbell, Hugh. 2009. 'Breaking New Ground in Food Regime Theory: Corporate Environmentalism, Ecological Feedbacks and the "Food from Somewhere" Regime?' *Agriculture and Human Values* 26 (4): 309–19.

Cardoso, Cíntia. 2002. 'Soja Transgênica Se Espalha Pelo Brasil'. *Folha de São Paulo*, 14 May.

Cardoso, Evorah Lusci, and Fabiola Fanti. 2013. 'Movimentos Sociais e Direito: O Poder Judiciário em Disputa'. In *Manual de Sociologia Jurídica*, edited by José Rodrigo Rodriguez and Felipe Gonçalves Silva, 237–54. São Paulo: Saraiva.

Carneiro, Fernando, Wanderlei Pignati, Raquel Maria Rigotto, L. G. S. Augusto, Anelise Rizollo, Neice Muller, Veruska Prado Alexandre, Karen Friedrich, and Marcia Sarpa de Mello. 2012. 'Agrotóxicos, Segurança Alimentar e Nutricional e e Saúde'. Vol. 1. Dossiê ABRASCO: Um Alerta Sobre os Impactos dos Agrotóxicos na Saúde. Rio de Janeiro: ABRASCO. www.abrasco.org.br/UserFiles/File/ABRASCODIVULGA/2012/DossieAGT.pdf.

Carrasco, Andrés. 2010. 'Carta Al CONICET Del Dr. Andrés Carrasco'Accessed 30 April 2013. www.lavaca.org/notas/fumiguen-a-la-ciencia/.

Carrizo, Cecilia Inés, and Mauricio Berger. 2009. *Estado Incivil y Ciudadanos sin Estado: Paradojas del Ejercicio de Derechos en Cuestiones Ambientales*. 1st ed. Córdoba, Argentina: Narvaja Ed.

'Carta do Rio Grande do Sul'. 1999. Accessed 30 July 2013. www.agirazul.com.br/artigos/cartars.htm.

Carter, Miguel. 2010. *Combatendo a Desigualdade Social: O MST e a Reforma Agrária no Brasil*. São Paulo: Ed. Unesp.

Castro, Bianca Scarpeline de. 2006. *O Processo de Institucionalização da Soja Transgênica no Brasil nos Anos de 2003 e 2005: A Partir da Perspectiva das Redes Sociais*. Master Thesis, Universidade Federal Rural do Rio de Janeiro.

Castro, Fabio de, and Renata Motta. 2015. 'Environmental Politics under Dilma: Changing Relations between the Civil Society and the State'. *LASA Forum* XLVI (3): 25–7.

Catacora-Vargas, Georgina, Pablo Galeano, Sarah Zanon Agapito-Tenfen, Darío Aranda, Tomás Palau, and Rubens Nodari. 2012. 'Report: Soybean Production in the Southern Cone of the Americas: Update on Land and Pesticide Use'. Cochabamba: Virmegraf. www.genok.com/news_cms/2012/july/report-soybean-production-in-the-southern-cone-of-the-americas-update-on-land-and-pesticide-use/158.

Cátedra Libre de Soberanía Alimentaria. 2013. 'Cómo Analizar La Nueva Ley de Semillas'. CaliSA, F.A.U.B.A., Buenos Aires.

Center for Global Development. 2014. 'MDG Progress Index: Gauging Country-Level Achievements'. *Center For Global Development*. Accessed 12 November 2014. www.cgdev.org/page/mdg-progress-index-gauging-country-level-achievements.

Cesarino, Letícia Maria Costa da Nóbrega. 2006. 'Acendendo as Luzes da Ciência Para Iluminar o Caminho do Progresso': Ensaio de Antropologia Simétrica da Lei de Biossegurança Brasileira'. Master Thesis, Universidade de Brasília.

Cintra, Lydia. 2013. 'A Natureza Reage às Monoculturas. É Algo Que Ela Considera Equivocado'. *Superinteressante*. Accessed 2 May 2014. http://super.abril.com.br/blogs/blogs/ideias-verdes/a-natureza-reage-as-monoculturas-e-algo-que-ela-considera-equivocado/.

CNBS. 2008a. 'Orientação CNBS No 1, de 31 de Julho de 2008'. Accessed 17 July 2011. www.ctnbio.gov.br/index.php/content/view/55.html?execview=listaitenslegislacao&norma=Orienta%E7%F5es.

——. 2008b. 'Orientação CNBS No 2, de 31 de Julho de 2008'. Accessed 17 July 2011. www.ctnbio.gov.br/index.php/content/view/55.html?execview=listaitenslegislacao&norma=Orienta%E7%F5es.

Comisión Nacional de Investigación Sobre Agroquímicos and CONICET. 2009. 'Evaluación de la Información Científica Vinculada al Glifosato en su Incidencia Sobre la Salud Humana y el Ambiente'. Buenos Aires. www.fundacion-campo. org/userfiles/prensa/glifosatoinfoconicet09.pdf.

Cordero, Mariano. 2001. 'En Bariloche los Comercios Deberán Identificar los Alimentos Transgénicos'. *El Clarín*, 10 May. http://edant.clarin.com/diario/2001 /05/10/s-03801.htm.

Córdoba, Maria Soledad. 2013. 'La Ruralidad Hiperconectada. Dinámicas de la Construcción de Redes en el Sector del Agro Argentino'. In *El Agro Como Negocio: Producción, Sociedad y Territorios en la Globalización*, edited by Carla Gras and Valeria Hernández, 263–88. Buenos Aires: Biblios.

Costa, Sérgio. 2006. *Dois Atlânticos*. Belo Horizonte: Editora UFMG.

Dąbrowska, Patrycja. 2007. 'Civil Society Involvement in the EU Regulations on GMOs: From the Design of a Participatory Garden to Growing Trees of European Public Debate'. *Journal of Civil Society* 3 (3): 287–304.

Das, Veena, and Deborah Poole, eds. 2004. *Anthropology in the Margins of the State*. Santa Fe, NM: School of American Research Press.

Delgado, Guilherme C. 2010. 'A Questão Agrária e o Agronegócio no Brasil'. In *Combatendo a Desigualdade Social: O MST e a Reforma Agrária no Brasil*, edited by Miguel Carter, 81–112. São Paulo: Ed. Unesp.

——. 2013. 'Pacto de Poder Com Os Donos da Terra'. *Le Monde Diplomatique Brasil*, 2 July, sec. www.diplomatique.org.br/artigo.php?id=1460.

Della Porta, Donatella, Hanspeter Kriesi, and Dieter Rucht, eds. 1999. *Social Movements in a Globalizing World*. Basingstoke: Macmillan.

Delvenne, Pierre, Federico Vasen, and Ana Maria Vara. 2013. 'The "Soy-ization" of Argentina: The Dynamics of the "Globalized" Privatization Regime in a Peripheral Context'. *Technology in Society* 35 (2): 153–62.

Desmarais, Annette Aurelie. 2007. *La Via Campesina: Globalization and the Power of Peasants*. London: Pluto Press.

'Dez Anos de Transgênicos no Brasil – Caminhos da Reportagem'. 2014. TVBrasil. www.youtube.com/watch?v=GbheATuAGbo&feature=youtube_gdata_player.

Diels, Johan, Cunha, Mario, Celia Manaia, Bernardo Sabugosa-Madeira, and Margarida Silva. 2011. 'Association of Financial or Professional Conflict of Interest to Research Outcomes on Health Risks or Nutritional Assessment Studies of Genetically Modified Products'. *Food Policy* 36: 197–203.

Doss, Cheryl, Gale Summerfield, and Dzodzi Tsikata. 2014. 'Land, Gender, and Food Security'. *Feminist Economics* 20 (1): 1–23.

Douglas, Mary. 2003. *Risk and Blame*. London: Routledge.

Dow AgroSciences. 2012. 'Dow AgroSciences Response to Activist Claims Regarding 2,4-D and Our New Herbicide-Tolerant Technology Package'. *Dow Agro-Sciences*. Accessed 30 April 2014. http://newsroom.dowagro.com/ press-release/dow-agrosciences-response-activist-claims-regarding-24-d- and-our-new-herbicide-toleran.

Edelman, Marc. 2008. 'Transnational Organizing in Agrarian Central America: Histories, Challenges, Prospects'. *Journal of Agrarian Change* 8 (2–3): 229–57.

Edelman, Marc, and Carwil James. 2011. 'Peasants' Rights and the UN System: Quixotic Struggle? Or Emancipatory Idea Whose Time Has Come?' *Journal of Peasant Studies* 38 (1): 81–108.

Editorial. 2001. 'MST Protesta Para Atrair Atenção No Fórum Social Mundial'. *Folha de São Paulo*, 26 January, print edition, sec. Cover.

Embrapa. 2014. 'Destino Do Agronegócio Brasileiro de Soja'. *Embrapa*. Accessed 3 March 2014. www.cnpso.embrapa.br/sojaemnumeros/app/graf6.html.

Embrapa Soja. 2003. 'Cronologia Do Embargo Judicial'. Accessed 1 March 2013. www.cnpso.embrapa.br/download/cronologia_sojarr.pdf.

Escobar, Arturo. 2001. 'Culture Sits in Places: Reflections on Globalism and Subaltern Strategies of Localization'. *Political Geography* 20 (2): 139–74.

——. 2008. *Territories of Difference: Place, Movements, Life, Redes*. Durham, NC: Duke University Press.

Ezcurra, Emiliano. 2001. 'Greenpeace No Juega'. *El Clarín*. http://edant.clarin.com/suplementos/rural/2001/02/10/r-00801.htm.

Fassi, Mariana C. 2009. 'Agricultura Empresarial y Globalizaciones. Los Efectos de La Soja Transgénica en el Paraguay'. *Revista Herramienta* 40. www.herramienta.com.ar/revista-herramienta-n-40/agricultura-empresarial-y-globalizaciones-los-efectos-de-la-soja-transgenic.

Fernandes, Gabriel. 2005. 'O Companheiro Liberou'. http://antigo.aspta.org.br/por-um-brasil-livre-de-transgenicos/legislacao-sobre-biosseguranca/o%20companheiro%20liberou.pdf.

——. 2010. 'Brief History of the GMFree Brazil Campaign'. In, 11. Rio de Janeiro. http://antigo.aspta.org.br/por-um-brasil-livre-de-transgenicos/updates/background-papers/.

Ferree, Myra Marx, William Anthony Gamson, Jürgen Gerhards, and Dieter Rucht. 2002. *Shaping Abortion Discourse: Democracy and the Public Sphere in Germany and the United States*. Cambridge: Cambridge University Press.

Figurelli, Maria. 2013. 'Movimientos Populares Agrarios: Asimetrías, Disputas y Entrelazamientos en la Construcción de Lo Campesino'. Working Paper Series 48. Berlin: desiguALdades.net International Research Network on Interdependent Inequalities in Latin America.

Fitting, Elizabeth M. 2011. *The Struggle for Maize: Campesinos, Workers, and Transgenic Corn in the Mexican Countryside*. Kindle version. Durham, NC: Duke University Press.

Folha de São Paulo. 2000a. 'Polêmica Divide Até Os Cientistas'. *Folha de São Paulo*, 4 March.

——. 2000b. 'Instituto Pede Medida Judicial Contra CNTBio'. *Folha de São Paulo*, 4 July.

——. 2000c. 'Greenpeace Faz Nova Lista de Transgênicos'. *Folha de São Paulo*, 20 September.

——. 2000d. 'Greenpeace Publica Lista de "Comida Segura"'. *Folha de São Paulo*, 5 October.

——. 2000e. 'Brasil Deve Influenciar Futuro Dos Transgênicos'. *Folha de São Paulo*, 17 May.

——. 2000f. 'Governo Sai Em Defesa Dos Transgênicos'. *Folha de São Paulo*, 7 July.

——. 2001a. 'Organização Pede Para ONU Apoiar Não-Transgênico'. *Folha de São Paulo*, 25 January.

——. 2001b. 'Idec Libera Lista de Alimentos Transgênicos'. *Folha de São Paulo*, 5 April.

——. 2001c. 'Indústrias Não Reconhecem Teste Do Idec'. *Folha de São Paulo*, 5 April.

——. 2001d. 'Greenpeace "Pinta" Transgênico Em Santa Cruz'. *Folha de São Paulo*, 5 July.

——. 2001e. 'Lula Volta a Criticar Liberação de Transgênicos'. *Folha de São Paulo*, 31 July.

——. 2001f. 'Greenpeace Divulga Nova Lista de Transgênicos Vendidos No Brasil'. *Folha de São Paulo*, 31 August.

——. 2001g. 'Prato Cheio'. *Folha de São Paulo*, 12 November.

——. 2001h. 'Idec Questiona o Decreto de Rotulagem'. *Folha de São Paulo*, 21 July.

——. 2002a. 'Análise Detecta Soja Modificada Em 5 Alimentos'. *Folha de São Paulo*, 16 March.

——. 2002b. 'Greenpeace Cria Guia de Produtos Com OGMs'. *Folha de São Paulo*, 30 May.

Foro de la Tierra y la Alimentación. 2002. 'Del "Granero del Mundo" a la Republiqueta Sojera: Por qué Estamos en contra del Modelo Transgénico'. Accessed 30 July 2012. www.greenpeace.org/argentina/Global/argentina/report/2006/3/del-granero-del-mundo-a-la.pdf.

Franco, Nádia. 2014. 'Ministério Diz Que é Contra Taxação de Exportações Agropecuárias'. *EBC Agência Brasil*, 31 March. http://agenciabrasil.ebc.com.br/economia/noticia/2014-03/ministerio-diz-que-e-contra-taxacao-de-exportacoes-agropecuarias.

Freitas, Amaury de Barros. 2011. 'Aliança Entre Movimentos Ambientalistas e de Consumidores: O Caso da Campanha por um Brasil Livre de Transgênicos'. Master Thesis, UFRRJ/CPDA.

Friedmann, Harriet, and Philip McMichael. 1989. 'Agriculture and the State System: The Rise and Decline of National Agricultures, 1870 to the Present'. *Sociologia Ruralis* 29 (2): 93–117.

Gamson, William A. 1992. *Talking Politics*. 5th ed. Cambridge: Cambridge University Press.

Gamson, William A., and David Meyer. 1996. 'Framing Political Opportunity'. In *Comparative Perspectives on Social Movements: Political Opportunities, Mobilizing Structures, and Cultural Framings*, edited by Doug McAdam, John D. McCarthy, and Mayer N. Zald, 275–90. Cambridge: Cambridge University Press.

Gamson, William A., and Andre Modigliani. 1989. 'Media Discourse and Public Opinion on Nuclear Power: A Constructionist Approach'. *The American Journal of Sociology* 95: 1–37.

Gaskell, George. 2004. 'Science Policy and Society: The British Debate over GM Agriculture'. *Current Opinion in Biotechnology* 15 (3): 241–5.

Gaskell, George, and Martin W. Bauer. 2001. *Biotechnology, 1996–2000: The Years of Controversy*. London: NMSI Trading Ltd.

Gerschmann, Léo. 1999. 'Governo Do RS Quer Banir Transgênico'. *Folha de São Paulo*, 9 March.

——. 2001. 'Uniao Liberará Venda de Soja Modificada'. *Folha de São Paulo*, 25 July.

Giarraca, Norma. 2001. 'El Movimiento de Mujeres Agropecuarias en Lucha: Protesta Agraria y Género Durante el último Lustro en Argentina'. In *Una Nueva Ruralidad en América Latina*, edited by Norma Giarracca, Edelmira C. Pérez, Maria de Nazareth Baudel Wanderley, and Miguel Teubal, 1st ed., 129–51. Buenos Aires: CLACSO.

Giarracca, Norma. 2003. 'De las Fincas y las Casas a las Rutas y las Plazas: Las Protestas y las Organizaciones Sociales en la Argentina de los Mundos "Rururbanos". Una Mirada Desde América Latina'. *Sociologias* 10: 250–83.

Giarracca, Norma, and Miguel Teubal. 2010. *Del Paro Agrario a Las Elecciones de 2009: Tramas, Reflexiones y Debates.* 1st ed. Buenos Aires: EA.

Giarracca, Norma, Edelmira C. Pérez, Maria de Nazareth Baudel Wanderley, and Miguel Teubal. 2001. *Una Nueva Ruralidad en América Latina.* Buenos Aires: CLACSO.

Girardi, E. P., and J. F. S. C. Vinha. 2013. 'Relatório Brasil 2012'. DATALUTA – Banco de Dados da Luta pela Terra. Presidente Prudente, São Paulo: NERA – Núcleo de Estudos, Pesquisas e Projetos de Reforma Agrária – FCT/ UNESP.

Giugni, Marco G. 1998. 'Was it Worth the Effort? The Outcomes and Consequences of Social Movements'. *Annual Review of Sociology* 24 (January): 371–93.

Globo Rural. 2013. 'Milho Resistente a Herbicida Causa Problema em Lavouras de Soja de MT'. http://g1.globo.com/economia/agronegocios/noticia/2013/12/milho-resistente-herbicida-causa-problema-em-lavouras-de-soja-de-mt.html.

Glover, Dominic. 2010. 'The Corporate Shaping of GM Crops as a Technology for the Poor'. *Journal of Peasant Studies* 37 (1): 67–90.

Gonçalves, Reinaldo. 2012. 'Novo Desenvolvimentismo e Liberalismo Enraizado'. *Serviço Social & Sociedade* 112 (December): 637–71.

Góngora-Mera, Manuel, and Renata C. Motta. 2014. 'El Derecho Internacional y la Mercantilización Biohegemónica de la Naturaleza: La Diseminación Normativa de la Propiedad Intelectual Sobre Semillas en Colombia y Argentina'. In *Desigualdades Socioambientales en América Latina,* edited by Barbara Göbel, Manuel Góngora-Mera, and Astrid Ulloa, 395–433. Bogotá: Universidad Nacional de Colómbia.

Goven, Joanna. 2003. 'Deploying the Consensus Conference in New Zealand: Democracy and De-Problematization'. *Public Understanding of Science* 12 (4): 423–40.

GRAIN. 2013a. 'La República Unida de La Soja Recargada'. *A Contrapelo,* 12 June.

——. 2013b. 'Leyes de Semillas en América Latina: Una Ofensiva Que No Cede y Una Resistencia Que Crece y Suma'. *A Contrapelo,* October.

Gramsci, Antonio. 1971. *Selections from the Prison Notebooks of Antonio Gramsci.* Edited and translated by Quintin Hoare and Geoffrey Nowel-Smith. London: Lawrence & Wishart.

Gras, Carla. 2009. 'Changing Patterns in Family Farming: The Case of the Pampa Region, Argentina'. *Journal of Agrarian Change* 9 (3): 345–64.

——. 2012. 'Cambio Agrario y Nueva Ruralidad: Caleidoscopio de la Expansión Sojera en la Región Pampeana'. *Trabajo y Sociedad* 18 (XIV): 7–24.

Gras, Carla, and Valeria Hernández. 2008. 'Modelo Productivo y Actores Sociales en el Agro Argentino'. *Revista Mexicana de Sociología* 70 (2): 227–59.

——, eds. 2013. *El Agro Como Negocio: Producción, Sociedad y Territorios en la Globalización.* Buenos Aires: Biblos.

Greenpeace Argentina. 2002. 'Cosecha Récord, Hambre Récord'. Accessed 30 July 2012. www.greenpeace.org/argentina/es/informes/la-guerra-de-estados-unidos-co/.

——. 2003. 'La Guerra de Estados Unidos contra el Sur y la Biodiversidad: Renovado ataque de un Estado agresivo'. Accessed 30 July 2012. www.greenpeace.org/argentina/es/informes/la-guerra-de-estados-unidos-co/.

——. 2013. 'Transgénicos'. Accessed 30 July 2012. www.greenpeace.org/argentina/es/campanas/bosques/transgenicos/.

Grupo de Madres de Córdova. 2005. 'Destrucción del Espacio Urbano: Genocidio Encubierto en Barrio Ituzaingó de Córdoba'. *Informe Alternativo Sobre la Salud en América Latina*. Quito, Ecuador: Observatorio Latinoamericano de la Salud CEAS. www.grr.org.ar/curitiba/sofiagatica. pdf.

Gudynas, Eduardo. 2008. 'The New Bonfire of Vanities: Soybean Cultivation and Globalization in South America'. *Development* 51 (4): 512–18.

——. 2012. 'Estado Compensador y Nuevos Extractivismos'. *Nueva Sociedad* 237: 128–46.

Guibu, Flávio. 2000. 'Manifestantes Depredam Navio da Libéria Com Milho Transgênico'. *Folha de São Paulo*, 26 July.

Guivant, Julia S. 2006. 'Transgênicos e Percepção Pública da Ciência No Brasil'. *Ambiente & Sociedade* 9 (1): 81–103.

Guivant, Julia S., and Philip Macnaghten. 2011. 'O Mito do Consenso: Uma Perspectiva Comparativa Sobre Governança Tecnológica'. *Ambiente & Sociedade* 14 (2): 89–104.

Habermas, Jürgen. 1992. 'Zur Rolle von Zivilgesellschaft und Politischer Öffentlichkeit'. In *Faktizität und Geltung: Beiträge zur Diskurstheorie des Rechts und des Demokratischen Rechtsstaats*, 4th ed., 399–467. Frankfurt am Main: Suhrkamp Verlag.

——. 2008. 'Hat die Demokratie Noch eine Epistemische Dimension? Empirische Forschung und Normative Theorie'. In *Ach, Europa: Kleine Politische Schriften XI*, 138–91. Frankfurt am Main: Suhrkamp Verlag.

Hafers, Luiz. 2000. 'A Discussão Sobre Transgênicos'. *Folha de São Paulo*, 22 August.

Hallin, Daniel C., and Paolo Mancini. 2004. *Comparing Media Systems: Three Models of Media and Politics*. Cambridge: Cambridge University Press.

Heller, Chaia. 2002. 'From Scientific Risk to Paysan Savoir-Faire: Peasant Expertise in the French and Global Debate over GM Crops'. *Science as Culture* 11 (1): 5–37.

——. 2013. *Food Solidarity: French Farmers and the Fight against Industrial Agriculture and Genetically Modified Crops*. Durham, NC: Duke University Press.

Hernández, Valeria. 2007. 'El Fenomeno Economico y Cultural del Boom de La Soja y el Empresariado Innovador'. *Desarrollo Económico* 47 (187): 331–65.

Herring, Ronald J. 2007. 'The Genomics Revolution and Development Studies: Science, Poverty and Politics'. *Journal of Development Studies* 43 (1): 1–30.

Hilson, Chris. 2002. 'New Social Movements: The Role of Legal Opportunity'. *Journal of European Public Policy* 9 (2): 238–55.

Hochstetler, Kathryn, and Margaret Elizabeth Keck. 2007. *Greening Brazil: Environmental Activism in State and Society*. Durham, NC: Duke University Press.

Holt Giménez, Eric, and Annie Shattuck. 2011. 'Food Crises, Food Regimes and Food Movements: Rumblings of Reform or Tides of Transformation?' *Journal of Peasant Studies* 38 (1): 109–44.

Hopkins, Terence K., and Immanuel Wallerstein. 1986. 'Commodity Chains in the World-Economy Prior to 1800'. *Review (Fernand Braudel Center)* 10 (1): 157–70.

Horst, Maja. 2010. 'Collective Closure? Public Debate as the Solution to Controversies about Science and Technology'. *Acta Sociologica* 53 (3): 195–211.

Idec. 2012. 'O Submundo Dos Agrotóxicos'. *Revista Do Idec*, December.

James, Clive. 2014. 'Global Status of Commercialized Biotech/GM Crops: 2013'. 46. ISAAA Brief. Ithaca, NY: ISAAA. www.isaaa.org/resources/publications/briefs/41/executivesummary/default.asp.

———. 2015. 'Global Status of Commercialized Biotech/GM Crops: 2014.' 49. ISAAA Briefs. Ithaca, NY: ISAAA. http://www.isaaa.org/resources/publications/briefs/49/

James, Clive, and A. F. Krattiger. 1996. 'Global Review of the Field Testing and Commercialization of Transgenic Plants, 1986 to 1995: The First Decade of Crop Biotechnology'. ISAAA Briefs. Ithaca, NY: ISAAA. www.isaaa.org/resources/publications/briefs/01/download/isaaa-brief-01-1996.pdf.

Jasanoff, Sheila. 2005. *Designs on Nature: Science and Democracy in Europe and the United States*. Princeton, NJ: Princeton University Press.

Jelin, Elizabeth, ed. 1990. *Women and Social Change in Latin America*. Women's Studies. Geneva: United Nations Research Inst. for Social Development.

———. 1996. 'Women, Gender, and Human Rights'. In *Constructing Democracy: Human Rights, Citizenship, and Society in Latin America*, edited by Elizabeth Jelin and Eric Hershberg, 177–96. Boulder, CO: Westview Press.

Johnson, Adrienne, and Anthony Bebbington. 2011. 'Rural Social Movements in Latin America: Organizing for Sustainable Livelihoods'. *Journal of Peasant Studies* 38 (3): 651–3.

Kaskey, Jack. 2013. 'Monsanto Offers Brazilian Farmers Royalty-Free Soybeans'. *Bloomberg*, 23 January. Accessed 23 January 2013. www.bloomberg.com/news/2013-01-23/monsanto-offers-brazilian-farmers-royalty-free-soybeans.html.

Kay, Cristóbal. 2001. 'Estructura Agraria, Conflicto y Violencia en la Sociedad Rural de América Latina'. *Revista Mexicana de Sociología* 63 (4): 159–95.

Keck, Margaret Elizabeth, and Kathryn. Sikkink. 1998. *Activists Beyond Borders: Advocacy Networks in International Politics*. Cambridge: Cambridge University Press.

Kerkvliet, Benedict J. Tria. 2009. 'Everyday Politics in Peasant Societies (and Ours)'. *Journal of Peasant Studies* 36 (1): 227–43.

Kinchy, Abby J. 2010. 'Epistemic Boomerang: Expert Policy Advice as Leverage in the Campaign Against Transgenic Maize in Mexico'. *Mobilization* 15 (2): 179–98.

———. 2012. *Seeds, Science, and Struggle: The Global Politics of Transgenic Crops*. Cambridge, MA: MIT Press.

Kleffmann & Partner SRL-Kleffmanngroup. 2013. 'Mercado Argentino de Produtos Fitosanitarios 2012'. www.casafe.org/pdf/estadisticas/Informe%20Mercado%20Fitosanitario%202012.pdf.

Kloppenburg, Jack Ralph. 2004. *First the Seed: The Political Economy of Plant Biotechnology, 1492 – 2000*. 2nd ed. Science and Technology in Society. Madison, WI: University of Wisconsin Press.

Koopmans, Ruud. 2005. 'The Missing Link Between Structure and Agency: Outline of an Evolutionary Approach to Social Movements'. *Mobilization* 10 (1): 19–33.

Lapegna, Pablo. 2013a. 'Notes From the Field: The Expansion of Transgenic Soybeans and the Killing of Indigenous Peasants in Argentina'. *Societies Without Borders* 8 (2): 291–308.

———. 2013b. 'Social Movements and Patronage Politics: Processes of Demobilization and Dual Pressure'. *Sociological Forum* 28 (4): 842–63.

——. 2014. 'Global Ethnography and Genetically Modified Crops in Argentina: On Adoptions, Resistances, and Adaptations'. *Journal of Contemporary Ethnography* 43 (2): 202–27.

——. 2015. 'Genetically Modified Soybeans, Agrochemical Exposure, and Everyday Forms of Peasant Collaboration in Argentina'. *The Journal of Peasant Studies* 0 (0): 1–20. doi:10.1080/03066150.2015.1041519.

Lavinas, Lena. 2012. 'Brasil, de la Reducción de la Pobreza al Compromiso de Erradicar la Miseria'. *Revista CIDOB D'afers Internacionals* 97–98 (April): 67–86.

——. 2013. '21st Century Welfare'. *New Left Review II* 84 (December): 5–40.

La Voz Del Interior. 2012. 'Nora Cortiñas Apoya a Las Madres de Ituzaingó', 11 June. www.lavoz.com.ar/ciudadanos/nora-cortinas-apoya-madres-ituzaingo.

Lazzarotto, Joelsio José, and Antônio Carlos Roessing. 2004. 'Contribuição da Agricultura Para a Arrecadação Tributária'. Embrapa Soja. www.contag.org.br/imagens/f1938embrapa-tributos.pdf.

Leguizamón, Amalia. 2014. 'Modifying Argentina: GM Soy and Socio-Environmental Change'. *Geoforum* 53 (May): 149–60.

Leite, Marcelo. 1998a. 'IDEC Quer Normas Mais Definidas'. *Folha de São Paulo*, 4 October.

——. 1998b. 'Para Empresa, Planta é Idêntica'. *Folha de São Paulo*, 4 October.

——. 1998c. 'SBPC Aguarda Esclarecimentos'. *Folha de São Paulo*, 4 October.

——. 1999. 'Campanha Defende Transgênicos'. *Folha de São Paulo*, 6 November.

——. 2000. 'ONG Testa Alimentos Em Busca de Ingredientes Transgênicos'. *Folha de São Paulo*, 15 February.

——. 2007. 'Arautos da Razão: A Paralisia no Debate Sobre Transgênicos e Meio Ambiente'. *Novos Estudos – CEBRAP* 78 (July): 41–7.

Levidow, Les, and Susan Carr. 1997. 'How Biotechnology Regulation Sets a Risk/Ethics Boundary'. *Agriculture and Human Values* 14 (1): 29–43.

Levidow, Les, Joseph Murphy, and Susan Carr. 2007. 'Recasting "Substantial Equivalence":Transatlantic Governance of GM Food'. *Science, Technology & Human Values* 32 (1): 26–64.

Lima, Dejoel de Barros. 2007. 'Legitimidade Social da Biotecnologia na Agricultura: O Caso da Soja Transgênica No Sul do Brasil'. PhD Thesis, Universidade Federal do Rio Grande do Sul.

Lisboa, Marijane. 2007. 'Transgênicos No Governo Lula: Liberdade Para Contaminar'. *PUCviva Revista* 29: 26–46.

——. 2009. 'Transgênicos: Quem Ganha Com Eles?' *PUCviva Revista* 36: 41–5.

Lopes, Reinaldo José, and Mari Tortato. 2006. 'País Defende Rastreamento de Transgênico'. *Folha de São Paulo*, 14 March.

Luhmann, Niklas. 2008. *Risk: A Sociological Theory*. New Brunswick, NJ: Aldine Transaction.

Lupion, Abelardo. 2005. 'Relatório Dos Trabalhos da CPMI da Terra'. Congresso Nacional. www.senado.gov.br/comissoes/CPI/RefAgraria/CPMITerra.pdf.

Luzzi, Mariana. 2012. *Problemas Socioeconómicos de La Argentina Contemporánea, 1976–2010*. Los Polvorines: Univ. Nacional de General Sarmiento.

Lyra, Letícia. 2005. 'A Mídia e o Debate Sobre Os Transgênicos'. Accessed 20 October 2013. www.portal-rp.com.br/pop/imprensa/2004_05_03.htm.

Magdoff, Fred, John Bellamy Foster, and Frederick H. Buttel, eds. 2000. *Hungry for Profit: The Agribusiness Threat to Farmers, Food, and the Environment*. New York: Monthly Review Press.

Magnan, André. 2006. 'Refeudalizing the Public Sphere: "Manipulated Publicity" in the Canadian Debate on GM Foods'. *The Canadian Journal of Sociology / Cahiers Canadiens de Sociologie* 31 (1): 25–53.

Marin, Anabel, Lilia Stubrin, and Patrick van Zwanenberg. 2014. 'Developing Capabilities in the Seed Industry: Which Direction to Follow?' Working Paper. Buenos Aires. http://media.wix.com/ugd/f051c2_f1171fa69237449a82e7ef95414 51e90.pdf.

Martínez Alier, Juan. 2003. *The Environmentalism of the Poor: A Study of Ecological Conflicts and Valuation.* Cheltenham: Edward Elgar.

Martínez-Torres, María Elena, and Peter M. Rosset. 2010. 'La Vía Campesina: The Birth and Evolution of a Transnational Social Movement'. *Journal of Peasant Studies* 37 (1): 149–75.

Mauersberger, Christof. 2015. *Advocacy Coalitions and Democratizing Media Reforms in Latin America: Whose Voice Gets on the Air?* Berlin: Springer.

Mayer, Sue, and Andy Stirling. 2004. 'GM Crops: Good or Bad?' *EMBO Reports* 5 (11): 1021–4.

McAdam, Doug, Sidney Tarrow, and Charles Tilly. 2001. *Dynamics of Contention.* Cambridge: Cambridge University Press.

McCann, Michael. 2006. 'Law and Social Movements: Contemporary Perspectives'. *Annual Review of Law and Social Science* 2 (1): 17–38.

McMichael, Philip. 2008. 'Peasants Make Their Own History, But Not Just as They Please …'. In *Transnational Agrarian Movements Confronting Globalization*, edited by Saturnino M. Borras, Marc Edelman, and Cristóbal Kay, 37–60. Chichester: Wiley-Blackwell.

——. 2009. 'A Food Regime Genealogy'. *Journal of Peasant Studies* 36 (1): 139–69.

Menasche, Renata. 2003. 'Os Grãos da Discórdia e o Risco à Mesa: Um Estudo Antropológico das Representações Sociais Sobre Cultivos e Alimentos Trans-gênicos No Rio Grande do Sul'. PhD Thesis, Universidade Federal do Rio Grande do Sul.

——. 2005. 'Os Grãos da Discórdia e o Trabalho da Mídia'. *Opinião Pública* 11 (1): 169–91.

Menezes, Thiago Melamed de, and Vicente Palermo. 2012. 'Gobierno de Lula y Lulismo: Examinando Algunas Hipótesis Sobre las Condiciones de Posibilidad y la Naturaleza del Lulismo'. *Temas y Debates* 23: 13–37.

Mesnage, R., E. Clair, S. Gress, C. Then, A. Székács, and G.-E Séralini. 2013. 'Cytotoxicity on Human Cells of Cry1Ab and Cry1Ac Bt Insecticidal Toxins Alone or with a Glyphosate-based Herbicide'. *Journal of Applied Toxicology* 33 (7): 695–9.

Mill, John Stuart. 1843. *A System of Logic, Ratiocinative and Inductive: Being a Connected View of the Principles of Evidence and the Methods of Scientific Investigation.* London: John W. Parker.

Ministerio de Agricultura, Ganadería y Pesca. 2010. 'Plan Estratégico Agroalimentario y Agroindustrial Participativo y Federal (PEA)'. www.produccioncatamarca.gov. ar/Publicaciones/files/33%20-%20Manual%20version%20final%20corregida%20 3%20de%20junio.pdf.

MOCASE-VC. 2012. '10 Razones Contra Modificaciones de Ley de Semillas En Argentina'. Blog do MOCASE Via Campesina – MNCI. Accessed 14 March 2013. http://mocase-vc.blogspot.com.ar/2012/10/10-razones-contra-modificaciones-de-ley.html.

Mohaded, Myriam. 2000. 'Una Carpa Negra Como Base de la Protesta Campesina'. *Página12*, 29 June. www.pagina12.com.ar/diario/elpais/1-123111-2009-04-13. html.

Motta, Renata. 2008. 'O Risco Nas Fronteiras Entre Política, Economia e Ciência: A Controvérsia Acerca da Política Sanitária Para Alimentos Geneticamente Modficados'. Master Thesis, Universidade de Brasília.

——. 2013a. 'Risky Politics: A Sociological Analysis of the WTO Panel on Biotechnological Products'. In *Balancing between Trade and Risk: Integrating Legal and Social Science Perspectives*, edited by Marjolein B.A. van Asselt, Esther Versluis, and Ellen Vos, 59–80. London and New York: Routledge.

——. 2013b. 'The Public Debate about Agrobiotechnology in Latin American Countries: A Comparative Study of Argentina, Brazil and Mexico'. 193. Production Development Series. Santiago de Chile: ECLAC.

——. 2014. 'Social Disputes over GMOs: An Overview'. *Sociology Compass* 8 (12): 1360–76.

——. 2015. 'Transnational Discursive Opportunities and Social Movement Risk Frames Opposing GMOs'. *Social Movement Studies* 14 (5): 576–95.

Motta, Renata, and Nadia Alasino. 2013. 'Medios y Política en la Argentina: Las Disputas Interpretativas Sobre la Soja Transgénica y el Glifosato'. *Question* 1 (38): 323–35.

Murakawa, Fabio Eduardo. 1999. 'Para Embrapa, Brasil Está Pronto Para Transgênicos'. *Folha de São Paulo*, 16 November.

Murillo, María Victoria, and Steven Levitsky. 2008. 'Argentina: From Kirchner to Kirchner'. *Journal of Democracy* 19 (2): 16–30.

Newell, Peter. 2006. 'Corporate Power and "Bounded Autonomy" in the Global Politics of Biotechnology'. In *The International Politics of Genetically Modified Food: Diplomacy, Trade, and Law*, edited by Robert Falkner, 67–85. Basingstoke: Palgrave Macmillan.

——. 2008. 'Trade and Biotechnology in Latin America: Democratization, Contestation and the Politics of Mobilization'. In *Transnational Agrarian Movements Confronting Globalization*, edited by Saturnino M. Borras, Marc Edelman, and Cristóbal Kay, 177–208. Chichester: Wiley-Blackwell.

——. 2009. 'Bio-Hegemony: The Political Economy of Agricultural Biotechnology in Argentina'. *Journal of Latin American Studies* 41 (1): 27–57.

Ninis, Alessandra Bortoni. 2011. *Complexidade, Manipulação Genética e Biocapitalismo: Compreensão das Interações da Engenharia Genética na Sociedade de Risco*. PhD Thesis, Universidade de Brasília.

Nobre, Marcos. 2013. *Imobilismo Em Movimento: Da Abertura Democrática Ao Governo Dilma*. São Paulo, SP: Companhia das Letras.

Nobre, Marcos, and José Rodrigo Rodriguez. 2011. '"Judicialização Da Política": Déficits Explicativos e Bloqueios Normativistas'. *Novos Estudos – CEBRAP* 91 (November): 5–20.

Otero, Gerardo, ed. 2008. *Food for the Few: Neoliberal Globalism and Biotechnology in Latin America*. Austin, TX: University of Texas Press.

'Otro Camino Para Superar La Crisis'. 2008. *Otro Camino Para Superar La Crisis*. Accessed 13 March 2014. http://otrocamino.wordpress.com/.

Paarlberg, Robert. 2000. 'Genetically Modified Crops in Developing Countries: Promise or Peril'. *Environment: Science and Policy for Sustainable Development* 42 (1): 19–27.

——. 2013. 'The World Needs Genetically Modified Foods'. *Wall Street Journal*, 14 April. http://online.wsj.com/article/SB10001424127887324105204578380872639718046.html.

Paganelli, Alejandra, Victoria Gnazzo, Helena Acosta, Silvia L. López, and Andrés E. Carrasco. 2012. 'Glyphosate-Based Herbicides Produce Teratogenic Effects on Vertebrates by Impairing Retinoic Acid Signaling'. *Chemical Research in Toxicology* 23 (10): 1586–95.

Palmer, Doug, and Robin Emmott. 2012. 'U.S. Trade Deal Could Be a Lot for Europe to Swallow'. *Reuters*, 11 December. Accessed 9 February 2014. www.reuters.com/article/2012/12/11/us-usa-eu-trade-idUSBRE8BA05Y20121211.

Pechlaner, Gabriela, and Gerardo Otero. 2008. 'The Third Food Regime: Neoliberal Globalism and Agricultural Biotechnology in North America'. *Sociologia Ruralis* 48 (4): 351–71.

Pelaez, Victor, and Wilson Schmidt. 2000. 'A Difusão dos OGM no Brasil: Imposição e Resistências'. *Estudos Sociedade e Agricultura* 14: 5–31.

Pellegrini, Pablo Ariel. 2013. *Transgénicos: Ciencia, Agricultura y Controversias en la Argentina*. Bernal: Universidad Nacional de Quilmes.

Pengue, Walter. 2005. 'Transgenic Crops in Argentina: The Ecological and Social Debt'. *Bulletin of Science, Technology & Society* 25 (4): 314–22.

Pereira, Paula, and Gabriel Fernandes. 2007. 'Mulheres Protestam Contra Milho Transgênico em Reunião da CTNBio'. 19 September. Accessed 10 March 2013. www.cptne2.org.br/index.php/publicacoes/noticias/noticias-do-campo/1734-mulheres-protestam-contra-milho-transgenico-em-reuniao-da-ctnbio.html.

Pérez Sáinz, Juan Pablo. 2015. '"Postneoliberalism" and Social Inequalities in the Andes: Reflections and Hypotheses on the Venezuelan, Bolivian, and Ecuadorian Cases'. In *A Moment of Equality for Latin America? Challenges for Redistribution*, edited by Barbara Fritz and Lena Lavinas, 53–76. Entangled Inequalities: Exploring Global Asymmetries. Burlington, VA: Ashgate.

Peters, Hans Peters. 2005. 'Editorial: Public Opinion on Biotechnology'. *International Journal of Public Opinion Research* 17 (1): 1–5.

Petras, James, and Henry Veltmeyer. 2001. 'Are Latin American Peasant Movements Still a Force for Change? Some New Paradigms Revisited'. *The Journal of Peasant Studies* 28 (2): 83–118.

Petras, James F., and Henry Veltmeyer. 2010. *Social Movements in Latin America: Neoliberalism and Popular Resistance*. Social Movements and Transformation. New York: Palgrave Macmillan.

Porto, Marcelo Firpo, and Bruno Milanez. 2009. 'Eixos de Desenvolvimento Econômico e Geração de Conflitos Socioambientais No Brasil: Desafios Para a Sustentabilidade e a Justiça Ambiental'. *Ciência & Saúde Coletiva* 14 (6): 1983–94.

Poth, Carla. 2013. 'Reconstruyendo la Institucionalidad del Modelo Biotecnológico Agrario: Un Enfoque Sobre la Comisión Nacional de Biotecnología Agropecuaria'. In *El Agro Como Negocio: Producción, Sociedad y Territorios en la Globalización*, edited by Carla Gras and Valeria Hernández, 289–322. Buenos Aires: Biblos.

Prevost, Gary, Carlos Oliva Campos, and Harry E. Vanden, eds. 2012. *Social Movements and Leftist Governments in Latin America: Confrontation or Co-Optation?* London; New York: Zed Books.

Przeworski, Adam, and Henry Teune. 1970. *The Logic of Comparative Social Inquiry.* New York: Wiley-Interscience.

Rauchecker, Markus. 2013. 'Intellectual Property Rights and Rent Appropriation – Open Conflict Regarding Royalties on RR Soy in Argentina'. *Journal für Entwicklungspolitik* 29 (2): 69–85.

——. 2015. 'Advocacy in Multi-Territorialen und Multi-Sektoralen Politischen Systemen – Der Wandel und die Konstanten der Pestizidregulierung im Fragmented State Argentinien'. PhD Thesis, Freie Universität Berlin.

REDAF. 2014. 'Conflictos por la Tierra y Ambientales'. *Red Agroforestal Chaco Argentina.* Accessed 13 January 2014. http://redaf.org.ar/observatorio/conflictos-por-la-tierra-y-ambientales/.

Reis, Maria Rita. 2012. 'Tecnologia Social de Produção de Sementes e Agrobiodiversidade'. Master Thesis, Universidade de Brasília.

Ribeiro, Isabelle Geoffroy, and Victor Augustus Marin. 2012. 'A Falta de Informação Sobre Os Organismos Geneticamente Modificados No Brasil'. *Ciência & Saúde Coletiva* 17 (2): 359–68.

Rihoux, Benoît, and Charles C. Ragin. 2009. *Configurational Comparative Methods: Qualitative Comparative Analysis (QCA) and Related Techniques.* Thousand Oaks, CA: SAGE.

Rippardo, Sergio. 2000. 'Oposição Quer Abrir CPI Dos Transgênicos'. *Folha de São Paulo,* 18 July.

Rippardo, Sergio, and Fabio Eduardo Murakawa. 2000. 'Pesquisa da Embrapa Favorece Monsanto'. *Folha de São Paulo,* 18 July.

Rodriguez, José Rodrigo. 2013. 'A Desintegração Do Status Quo: Direito e Lutas Sociais'. *Novos Estudos – CEBRAP* 96 (July): 49–66.

Rucht, Dieter. 1994. 'Öffentlichkeit Als Mobilisierungsfaktor für Soziale Bewegungen'. *Kölner Zeitschrift für Soziologie und Sozialpsychologie* 34 (Sonderhefte): 337–58.

Rulli, Jorge Eduardo. 2009. *Pueblos Fumigados: Los Efectos de los Plaguicidas en las Regiones Sojeras.* Buenos Aires: Del nuevo extremo.

Salazar, Andrea Lazzarini. 2011. 'A Informação Sobre Alimentos Transgênicos No Brasil'. In *Transgênicos Para Quem? Agricultura, Ciência, Sociedade,* edited by Magda Zagnoni and Gilles Ferment, 302–16. Brasília: Ministério do Desenvolvimento Agrário.

Salazar, Andrea Lazzarini, and Katarina Bozola Grou. 2010. 'Alimentos Transgênicos, Direitos Humanos e o Poder Judiciário'. In *Justiça e Direitos Humanos: Experiências de Assessoria Jurídica Popular,* edited by Darci Frigo, Fernando Prioste, and Antônio Sérgio Escrivão Filho, 85–103. Curitiba: Terra de Direitos.

Samora, Roberto. 1999. 'MST Anuncia Invasao de Fazendas Com Transgênicos'. *Folha de São Paulo,* 19 December.

Sampaio Jr, Plínio de Arruda. 2012. 'Desenvolvimentismo e Neodesenvolvimentismo: Tragédia e Farsa'. *Serviço Social & Sociedade* 112 (December): 672–88.

Santos, Laymert Garcia dos. 2007. 'Desencontro ou "Malencontro"? Os Biotecnólogos Brasileiros em Face da Sócio e da Biodiversidade'. *Novos Estudos – CEBRAP* 78 (July): 49–57.

Sauer, Sérgio. 2008. 'Agricultura Familiar versus Agronegócio: A Dinâmica Sociopolítica do Campo Brasileiro'. Textos para Discussão 30. Brasília: Embrapa Informação Tecnológica.

——. 2010. *Terra e Modernidade: A Reinvenção do Campo Brasileiro*. São Paulo: Expressão Popular.

Scheinberg, Gabriela. 2000. 'Campanha Pede Mais Rigor Sobre Transgenicos'. *Folha de São Paulo*, 2 February.

Schild, Verónica. 1994. 'Recasting "Popular" Movements: Gender and Political Learning in Neighborhood Organizations in Chile'. *Latin American Perspectives* April (21): 59–80.

——. 2013. 'Care and Punishment in Latin America: The Gendered Neoliberalisation of the Chilean State'. In *Neoliberalism, Interrupted: Social Change and Contested Governance in Contemporary Latin America*, edited by Mark Goodale and Nancy Postero, 195–224. Stanford: Stanford University Press.

Schnurr, Matthew A. 2013. 'Biotechnology and Bio-Hegemony in Uganda: Unraveling the Social Relations Underpinning the Promotion of Genetically Modified Crops into New African Markets'. *Journal of Peasant Studies* 40 (4): 639–58.

Schurman, Rachel, and William Munro. 2009. 'Targeting Capital: A Cultural Economy Approach to Understanding the Efficacy of Two Anti-Genetic Engineering Movements'. *American Journal of Sociology* 115 (1): 155–202.

——. 2010. *Fighting for the Future of Food: Activists versus Agribusiness in the Struggle over Biotechnology*. Social Movements, Protest, and Contention 35. Minneapolis, MN: University of Minnesota Press.

Scoones, Ian. 2008. 'Mobilizing against GM Crops in India, South Africa and Brazil'. In *Transnational Agrarian Movements Confronting Globalization*, edited by Saturnino M. Borras, Marc Edelman, and Cristóbal Kay, 147–76. Chichester: Wiley-Blackwell.

Scott, James C. 1987. *Weapons of the Weak: Everyday Forms of Peasant Resistance*. New Haven, CT: Yale University Press.

Secretaría de Agricultura, Ganadería, Pesca y Alimentos de la Nación. 2004. 'Plan Estratégico 2005, 2015 Para el Desarrollo de la Biotecnología Agropecuaria'. www.argenbio.org/adc/uploads/pdf/Plan2015.pdf.

Seifert, Franz. 2011. 'Back to Politics at Last – Orthodox Inertia in the Transatlantic Conflict over Agro-Biotechnology'. *Science, Technology & Innovation Studies* 6 (2): 101–26.

Seifert, Roberto. 2012. 'La Nueva Ley de Semillas Sacudió Todo el Espectro Rural'. *La Nácion*, 29 September. www.lanacion.com.ar/1512410-la-nueva-ley-de-semillas-sacudio-todo-el-espectro-rural.

Seoane, José, Emilio Taddei, and Clara Algranati. 2006. 'Las Nuevas Configuraciones de Los Movimentos Populares en América Latina'. In *Política y Movimientos Sociales en un Mundo Hegemónico Lecciones Desde África, Asia y América Latina*, edited by Atilio A. Boron and Gladys Lechini, 227–50. Buenos Aires: CLACSO.

Séralini, Gilles-Eric, Emilie Clair, Robin Mesnage, Steeve Gress, Nicolas Defarge, Manuela Malatesta, Didier Hennequin, and Joël Spiroux de Vendômois. 2012. 'Long Term Toxicity of a Roundup Herbicide and a Roundup-Tolerant Genetically Modified Maize'. *Food and Chemical Toxicology* 50 (11): 4221–31.

Séralini, G. E., R. Mesnage, N. Defarge, and J. Spiroux de Vendômois. 2014. 'Conclusiveness of Toxicity Data and Double Standards'. *Food and Chemical Toxicology* 69 (July): 357–9.

Shefner, Jon. 2004. 'Introduction: Current Trends in Latin American Social Movements'. *Mobilization* 9 (3): 219–22.

Sigaud, Lygia, Marcelo Rosa, and Marcelo Ernandez Macedo. 2008. 'Ocupações de terra, acampamentos e demandas ao Estado: uma análise em perspectiva comparada'. *Dados* 51 (1): 107–42.

Silva, Eduardo. 2009. *Challenging Neoliberalism in Latin America*. Cambridge Studies in Contentious Politics. Cambridge: Cambridge University Press.

Silva, Eliana. 2000. 'Reuniao do MST Prega Invasão e Queimada'. *Folha de São Paulo*, 12 August.

Silveira, Cristiane Amaro da. 2004. 'Significados Sociais das Biotecnologias: Interesses e Disputas em Torno dos Organismos Geneticamente Modificados (OGMs) no Rio Grande do Sul'. Master Thesis, Universidade Federal do Rio Grande do Sul.

Silveira, Cristiane Amaro da, and Jalcione Almeida. 2008. 'Tecnociência, Democracia e Os Desafios éticos das Biotecnologias No Brasil'. *Sociologias* 10 (19): 106–29.

Singer, André Vítor. 2012. *Os Sentidos do Lulismo: Reforma Gradual e Pacto Conservador*. São Paulo: Companhia das Letras.

Skocpol, Theda. 1976. 'France, Russia, China: A Structural Analysis of Social Revolutions'. *Comparative Studies in Society and History* 18 (2): 175–210.

Skocpol, Theda, and Margaret Somers. 1980. 'The Uses of Comparative History in Macrosocial Inquiry'. *Comparative Studies in Society and History* 22 (2): 174–97.

Snow, David A., and Robert D. Benford. 1988. 'Ideology, Frame Resonance, and Participant Mobilization'. *International Social Movement Research* 1: 197–217.

——. 1992. 'Master Frames and Cycles of Protest'. In *Frontiers in Social Movement Theory*, edited by Aldon D. Morris and Carol McClurg Mueller, 133–55. New Haven, CT and London: Yale University Press.

Snow, David A., E. Burke Rochford., Steven K. Worden, and Robert D. Benford. 1986. 'Frame Alignment Processes, Micromobilization, and Movement Participation'. *American Sociological Review* 51: 464–81.

Sociedade Rural Brasileira and Embrapa. 1999. 'A Importância da Biotecnologia Para o Brasil'. *Folha de São Paulo*, 4 November.

Sotomayor, Octavio, Adrián Rodríguez, and Mônica Rodrigues. 2011. *Competitividad, Sostenibilidad e Inclusión Social en la Agricultura. Nuevas Direcciones en el Diseño de Políticas en América Latina y El Caribe*. Santiago de Chile: Naciones Unidas.

Stanley, Jason, and Jeff Goodwin. 2013. 'Political Economy and Social Movements'. In *The Wiley-Blackwell Encyclopedia of Social and Political Movements*, 946–49. Blackwell Publishing Ltd.

Streeck, Wolfgang. 2011. 'The Crises of Democratic Capitalism'. *New Left Review II* 71 (October): 5–29.

Svampa, Maristella. 2008. *Cambio de Época: Movimientos Sociales y Poder Político*. Buenos Aires: Clacso-Siglo XXI Editores Argentina.

——. 2012. 'Consenso de los Commodities, Giro Ecoterritorial y Pensamiento Crítico en América Latina'. *Revista Del Observatorio Social de América Latina* XIII (32).

Taglialegna, Gustavo Henrique F. 2005. 'Grupos de Pressão e a Tramitação do Projeto de Lei de Biossegurança No Congresso Nacional'. Senado Federal. http://www2. senado.gov.br/bdsf/bitstream/id/110/4/texto28%20-%20gustavo.pdf.

Tarrow, Sidney G. 2005. *The New Transnational Activism*. New York: Cambridge University Press.

——. 2011. *Power in Movement: Social Movements and Contentious Politics*. Kindle edition. Cambridge Studies in Comparative Politics. Cambridge: Cambridge University Press.

Tesh, Sylvia N. 2000. *Uncertain Hazards: Environmental Activists and Scientific Proof*. 1st ed. Ithaca, NY: Cornell University Press.

Teubal, Miguel, and Javier Rodríguez. 2002. *Agro y Alimentos en la Globalización: Una Perspectiva Crítica*. Colección Agricultura y Ciencias Sociales. Buenos Aires: Ed. La Colmena.

The Africa Centre for Biosafety. 2013. 'GM Industry Called to Account: ISAAA's Report Mischievous and Erroneous'. *The Africa Centre for Biosafety*. Accessed 27 February 2013 www.acbio.org.za/index.php/media/64-media-releases/418.

The Economist. 2010. 'Economist Debates: Biotechnology'. *The Economist*. Accessed 12 March 2013. www.economist.com/debate/overview/187.

Thuswohl, Maurício. 2013. 'Na Câmara, Proposta Tenta Liberar Sementes Transgênicas "Suicidas"'. *Repórter Brazil*. 11 November. Accessed 3 April 2014. http:// reporterbrasil.org.br/transgenicos/na-camara-proposta-tenta-liberar-sementes-transgenicas-suicidas/.

Tilly, Charles, and Sidney Tarrow. 2007. *Contentious Politics*. Boulder, CO: Paradigm Publishers.

Todoagro.com.ar. 2013. 'Definen las Pautas Sobre Aplicaciones de Productos Fitosanitarios en áreas Periurbanas'. *Todoagro.com.ar*. 22 October. Accessed 23 October 2013. www.todoagro.com.ar/noticias/nota.asp?nid=26087.

Tortato, Mari. 2003. 'Requião Fecha Paranaguá Para Transgênico'. *Folha de São Paulo*, 28 October. http://www1.folha.uol.com.br/fsp/dinheiro/fi2810200324. htm.

——. 2006. 'País Defende Rastreamento de Transgênico'. *Folha de São Paulo*, 14 March.

Traumann, Thomas. 2001. 'Monsanto Deve Processar Bové e Stédile'. *Folha de São Paulo*, 29 January.

Trigo, Eduardo J., Daniel Chudnovsky, and Eugenio Cap. 2002. *Los Transgénicos en la Agricultura Argentina: Una Historia con Final Abierto*. Buenos Aires: Libros del Zorzal.

Trucco, Victor. 2001. 'El Juego de Greenpeace'. *El Clarín*, 3 February. http://edant. clarin.com/suplementos/rural/2001/02/03/r-00801.htm.

UNCTAD. 2015. 'The State of Commodity Dependence 2014'. New York and Geneva: UNCTAD. http://unctad.org/en/PublicationsLibrary/suc2014d7_en.pdf.

UPC- UNESCO, EdPAC, GCCT, and GIDHS. 2009. 'Situación de Los Derechos Humanos en el Noroeste Argentino en 2008'. Barcelona: Càtedra Unesco en Sostenibilitat. http://edpac.org/docs/Publicacio_Informe_Argentina.pdf.

Vaccarezza, Cândido. 2014. 'Em Defesa da Pesquisa'. *Folha de São Paulo*, 16 January. http://www1.folha.uol.com.br/opiniao/2014/01/1398224-candido-vaccarezza-em-defesa-da-pesquisa.shtml.

Valdés, Maria Fernanda. forthcoming. *Reducing Inequality in Latin America: The Role of Tax Policy*. Aldershot: Ashgate.

Valente, Rubens. 2003. 'Lobby Transgênico Leva Deputados Aos EUA'. *Folha de São Paulo*, 14 June Poder. http://www1.folha.uol.com.br/folha/brasil/ult96u50132. shtml.

Van den Daele, Wolfgang. 2007. 'Legal Framework and Political Strategy in Dealing with Risks of New Technology: The Two Faces of the Precautionary Principle.' In *Regulatory Challenge of Biotechnology: Human Genetics, Food and Patents*, edited by Han Somsen, 118–38. Cheltenham, UK; Northampton, USA: Edward Elgar.

van Zwanenberg, Patrick, and Valeria Arza. 2013. 'Biotechnology and Its Configurations: GM Cotton Production on Large and Small Farms in Argentina'. *Technology in Society* 35 (2): 105–17.

Varesi, Gastón Ángel. 2010. 'El Circuito Productivo Sojero Argentino en el Modelo Posconvertibilidad. Una Aproximación Desde el Enfoque de Análisis Regional'. *Cuadernos Del CENDES* 74: 107–40.

Veja. 2011. 'Brasil Bate Produção de Soja Argentina, Mas Ganha Menos', 11 September. Accessed 2 March 2014. http://veja.abril.com.br/noticia/economia/pais-bate-producao-de-soja-argentina-mas-ganha-menos.

Veltri, Giuseppe A., and Ahmet K. Suerdem. 2013. 'Worldviews and Discursive Construction of GMO-Related Risk Perceptions in Turkey'. *Public Understanding of Science* 22 (2): 137–54.

Via Campesina. 2006. 'La Via Campesina Occupies an Area Planted with Illegal Transgenic Seeds in Paraná'. *La Via Campesina*. 15 March. Accessed 19 April 2014. http://viacampesina.org/en/index.php/main-issues-mainmenu-27/biodiversity-and-genetic-resources-mainmenu-37/103-la-via-campesina-occupies-an-area-planted-with-illegal-transgenic-seeds-in-paran.

Voces de Alerta. 2009. 'Voces de Alerta'. 23 May. Accessed 14 March 2014. http://voces-de-alerta.blogspot.de/2009_05_01_archive.html.

Wagner, Peter. 2013. 'Provinz und Welt. Demokratie und Kapitalismus in Europa, Brasilien und Südafrika'. *Westend* 1: 38–60.

Watts, Jonathan. 2013. 'Unease among Brazil's Farmers as Congress Votes on GM Terminator Seeds'. *The Guardian*, 12 December. Accessed 15 December 2013. www.theguardian.com/global-development/2013/dec/12/brazil-gm-terminator-seed-technology-farmers.

World Trade Organization. 2006. 'EC – Approval and Marketing of Biotech Products'. Accessed 30 April 2014. www.wto.org/english/tratop_e/dispu_e/cases_e/ds293_e.htm.

Wylde, Christopher. 2010. '*Argentina, Kirchnerismo, and Neodesarrollismo: Argentine Development under Néstor Kirchner, 2003–2007*'. Documento de Trabajo n. 44. Buenos Aires: FLACSO.

Wynne, Brian. 2001. 'Creating Public Alienation: Expert Cultures of Risk and Ethics on GMOs'. *Science as Culture* 10 (4): 445–81.

Yano, Célio. 2009. 'Transgênico 100% Nacional'. *Ciência Hoje*. 23 September. Accessed 20 July 2013. http://cienciahoje.uol.com.br/noticias/2011/09/transgenico-100-nacional.

Zanatta, Mauro. 2009. 'Embrapa Manifesta-Se Contra o Plantio de Arroz Transgênico'. *Valor Econômico*, 19 March.

Zhouri, Andréa, and Raquel Oliveira. 2012. 'Development and Environmental Conflicts in Brazil: Challenges for Anthropology and Anthropologists'. *Vibrant: Virtual Brazilian Anthropology* 9 (1): 181–208.

Index